FAILURE TO COMMUNICATE

WHY WE MISUNDERSTAND WHAT WE HEAR, READ, AND SEE

Roger Kreuz

Prometheus Books

Essex, Connecticut

Prometheus Books

An imprint of Globe Pequot, the trade division of The Rowman & Littlefield
Publishing Group, Inc.
4501 Forbes Blvd., Ste. 200
Lanham, MD 20706
www.rowman.com

Distributed by NATIONAL BOOK NETWORK

British Library Cataloguing in Publication Information Available

Library of Congress Cataloging-in-Publication Data Available

ISBN 9781633888890 (cloth : alk. paper) | ISBN 9781633888906 (epub)

♾™ The paper used in this publication meets the minimum requirements of
American National Standard for Information Sciences—Permanence of Paper
for Printed Library Materials, ANSI/NISO Z39.48-1992

For Bob and Sheila

It was not death she feared. It was misunderstanding.

—Zora Neale Hurston,
Their Eyes Were Watching God (1937)

"Most of the trouble in life comes from misunderstanding,
I think," said Anne.

—L. M. Montgomery,
Anne of the Island (1915)

CONTENTS

PROLOGUE

We live in a time when we have more options for communication than ever before. But are we communicating any better? In many ways, it seems that we aren't. In terms of communicative choice, it is the best of times. We can text, blog, post, share, and tweet with abandon—and many people do. But in terms of actual communication, it may well be the worst of times. Many of these new mediums are pale analogues of real interactions, and they can lead to all sorts of misunderstandings. We might receive a tweet or a text in an instant but have little idea about what its sender really intends. Is the message meant to be serious or playful? Is its brevity a sign that its sender is annoyed—or just in a hurry? And what does its punctuation—or lack thereof—mean? We might find ourselves making all sorts of assumptions that are unwarranted and inferences that are incorrect.

We stand at a moment in history when communicating seems to be more fraught than ever. One reason for this is that these new ways to communicate are proliferating and evolving rapidly. We may struggle, for example, to compose a business email that we intend to be light-hearted and then find ourselves uncertain about whether our intent is clear. Perhaps an emoji would help? But which one? And would this be unprofessional in a business context?

In the 1980s, when online bulletin boards, chat rooms, and news-groups were all the rage, the term "netiquette" was coined to refer to informal codes of practice for using them. But since these standards were informal, not everyone was aware of them or followed them. We find ourselves in much the same situation today, but with a plethora of new ways to communicate and with just as little agreement about how to do it.

Let's take a step back and imagine an ideal scenario for effective communication: joining a good friend for coffee in a nearly deserted café. It's easy to converse because everything is working in your favor. You can see and easily hear your conversational partner, which means that a variety of verbal and nonverbal cues are available to aid your comprehension. Her tone, facial expressions, and even posture provide you with a rich set of signals for understanding what she is saying. And because you know her well, you know that when she cocks her head to the side, she's about to make a joke or a sly observation. Understanding what she is saying—and what she means—seems almost effortless.

But now let's complicate things a bit. The café, which had been almost empty, begins to fill up with noisy patrons. You find yourself straining a bit to hear her over the din. And then she makes reference to a person you've never heard her mention before. Who is she talking about? You might need to interrupt to ask for a repetition or clarification. What had been so easy and comfortable a few moments ago now seems more effortful.

Let's increase the level of difficulty once again. Instead of meeting a good friend at the noisy café, imagine that you're on a blind date. You know almost nothing about the other person except that a friend of a friend thinks you might enjoy getting to know each other. You find yourself making small talk, but it's forced and awkward. What topics are safe to discuss? What does it mean when she rolls her eyes in mid-sentence? Was she making a joke about the service, or was she being serious? And why is everybody talking so loudly? Such a conversation would be a lot of work and probably not particularly enjoyable. In addition, opportunities for miscommunication would be legion.

In this book, I will argue that communication can be thought of as robust but also brittle. It is robust because it can typically withstand one source of degradation, disruption, or ambiguity. However, it is brittle because two or more types of disruption can cause communication to break down. A noisy café makes it more difficult to hear your friend, but you can manage until she refers to someone who is unknown to you. And interacting with a complete stranger in a noisy environment would be an exercise in patience and goodwill for both of you. And what if your blind date were taking place over a bad cellphone connection instead of face-to-face? With all these factors working against you, it's unlikely that you would have an especially meaningful conversation. Whenever there are two or more impediments to communication, it becomes clear how fragile and prone to breaking the process of understanding can be.

In an important sense, humanity and communication evolved together. For thousands of years, people interacted face-to-face with others who they knew well. This has changed only recently. Technology has, at one level, made communication easier than ever. Writing and widespread literacy enabled us to exchange letters with others who might be far away. The telephone allowed us to communicate in real time with friends and loved ones. Electronic mail made it possible to send missives throughout the world in the blink of an eye. And text messaging, tweets, and social media posts have taken communication to a whole new level. Despite all of this, our ability to communicate effectively has not increased and may in fact have gotten worse.

Telephone calls, unless they are video chats, strip away many of the nonverbal cues that help us understand the meaning of what other people say. Email is even more problematic since it eliminates both the verbal and the nonverbal cues that can facilitate comprehension. And social media posts by strangers can be almost impossible to interpret. Is the poster being serious or mischievous? Whenever we move away from face-to-face conversations with intimates, we put stress on the communicative process. And if we bend it far enough, it will break.

As we reflect on our communicative misadventures, we might find ourselves trying to figure out why they occurred. And, as with many things in life, we are likely to arrive at the conclusion that the fault must lie elsewhere. We might, for example, decide that a misunderstanding occurred because of another person's inattention or lack of knowledge. Social psychologists refer to such reflexive judgments as examples of correspondence bias. We tend to view the failures of others as arising from their personal qualities, whereas our own failings are the result of situational factors. If we return to the café example described earlier, we can see how this might play out. You might conclude that your conversational partner failed to understand something you said because she was distracted, which may not be out of character for her. However, you might attribute your failure to understand *her* to the din caused by the other patrons in the café.

What many people fail to realize is that there are a host of additional reasons why communication sometimes goes off the rails. An important one is the language itself. A language like English was not designed to allow for clear and unambiguous communication. And that's because it wasn't designed at all. Instead, it gradually acquired its modern form through a series of historical accidents that affected its sound structure, its grammar, and its vocabulary. It grew organically, not beholden to the rules of logic. Parts of it, like the spelling of words, have become frozen, even as other aspects, such as how we pronounce those words, continue to change.

In addition, the ways that our bodies and minds work affect the process of communication. Some causes of misunderstanding are sensory and are the result of how our sense organs function. Others are perceptual and arise from how our minds interpret what we see and hear. And still other communicative problems have a cognitive basis and are tied to memory and thought. Finally, as we have already seen, social and cultural factors play an important role as well. As it turns out, there is plenty of blame to go around.

Fortunately, we know a great deal about how communication works. In the past several decades, psychologists, linguists, and other researchers have created a multidisciplinary framework that has become known as cognitive science. Researchers working within this theoretical orientation characterize communication in much the way that I've described it here: as a set of interlocking perceptual, cognitive, linguistic, and social factors. And as with many other complex processes, all it takes is one or two problems to trigger a systemic failure. Communication is remarkably robust, but it is far from bulletproof. And being aware of the sources of miscommunication is the first step to communicating more effectively.

In this book, I will make use of this cognitive science perspective—and the volumes of research it has generated—to describe and explain the issues that can arise when people speak, hear, write, and read. I'll ground this discussion in dozens of real-world examples of communication failure. My goal is to explain how and why miscommunication occurs in a wide variety of contexts. By the end, you'll have a better understanding of the strengths—and also the fragility—of the communicative process.

ACKNOWLEDGMENTS

It may not take a village to write a book, but it does require the assistance, the patience, and the goodwill of many. I'm grateful to everyone who helped make this project a reality.

I'm especially thankful to my beta readers, who were willing to pore over the first draft of this book and to provide invaluable feedback. You should be thankful as well: they suffered so that you wouldn't have to. Richard, Rick, and Katherine—I'm in your debt. The comments of two anonymous reviewers were also very helpful.

The services of the interlibrary loan department of the Ned McWherter Library at the University of Memphis were invaluable. No matter how obscure my request, their crackerjack staff was always able to find what I needed and to send it to me quickly. Librarians are truly the unsung heroes behind a book like this one.

I must also thank my dean, Abby Parrill; my department chair, Randy Floyd; and the administrative staff at Scates Hall, especially Tori Tardugno, for their support and encouragement.

At Prometheus Books, I'm very grateful to Jon Kurtz for his initial interest in my proposal and his willingness to take this project on when it returned to him. Thanks also go to the sharp eyes of Bruce Owens, my copyeditor, and Nicole Carty, my production editor.

Finally, I'm very grateful to have an agent like Andy Ross, who has been indefatigable in his efforts on my behalf.

Roger Kreuz
Memphis, Tennessee
March 2022

1

ASPECTS OF MISCOMMUNICATION

To begin our discussion, let's consider some real-world examples of miscommunication. In this chapter, we will see how such misunderstandings are caused by the assumptions and expectations that we bring to our interactions with others. To a great extent, we aren't even aware of these factors—that is, until they lead to some sort of communicative failure. Sometimes the problems that arise are trivial, but as we will see, some have been consequential.

"PASADENA, WE'VE GOT A PROBLEM"

On December 11, 1998, the Mars Climate Orbiter roared off the launchpad at Cape Canaveral to begin its journey to the Red Planet. The unmanned orbiter was billed as the "first interplanetary weather satellite,"[1] designed to study the atmosphere and meteorological conditions of Mars. In addition, the craft would serve as a communications link for the Mars Polar Lander, which would arrive at Mars the following year.[2] The orbiter, which had cost $285 million to build and launch, operated normally throughout its long flight.

Ten months later, on September 23, 1999, the spacecraft began its Martian orbital insertion maneuver by firing rockets to slow its speed. There was an anticipated loss of radio contact as it flew around the far side of the planet, and the flight controllers waited anxiously for a sign from the spacecraft after emerging from this radio blackout. As the minutes dragged by without a signal, the technicians at the Jet Propulsion Laboratory (JPL) in Pasadena, California, had their worst fears realized: something had gone terribly wrong. The orbiter was never heard from again.[3]

A subsequent investigation was able to pinpoint the cause of the mission failure. The craft's trajectory had taken it too close to the surface of the planet, where it had either burned up or skipped off the Martian atmosphere and back into space. This happened because Lockheed Martin, the contractor responsible for designing and building the spacecraft, wrote thruster software that measured impulse in terms of pound seconds. NASA's guidance system, however, made use of newton seconds as a unit of force instead. The resulting discrepancy led to a significant error in the trajectory calculation, causing the loss of the spacecraft and the mission.

Lockheed Martin, like the rest of the U. S. launch industry, had made use of so-called customary units, such as feet and pounds, derived from the British Imperial system. The United States is one of only three countries that do not currently use metric system units like meters and kilograms (the other two are Liberia and Myanmar). The engineers at JPL had assumed that any necessary conversions had been made, but this was not the case. The failure to detect the discrepancy in units of measurement, as well as the subsequent loss of the Mars Polar Lander spacecraft, caused NASA to rethink the "faster, better, and cheaper" approach that had been the agency's mantra during the 1990s.[4]

Although the loss of the Mars Climate Orbiter provides a particularly poignant (and costly) example, misaligned frames of reference can lead to miscommunication and misunderstandings in a wide variety of contexts. Such misalignments are perhaps most obvious when they involve cultural groups with different beliefs or expectations. The historical record is replete with episodes of two parties talking past one another. This can be clearly seen in the case of treaties negotiated between colonial settlers and the native peoples of North America.

Generations of U.S. schoolchildren, for example, have learned that in 1626, Peter Minuit purchased the island of Manhattan from the Lenape tribe. In return for the land, the Dutch governor gave the Lenape goods worth about 60 guilder. The items themselves went unrecorded but were probably trade goods, such as cloth, and metal items, such as tools.[5] (This was equivalent to about $24 when an account of the transaction was popularized in the mid-nineteenth century.[6] Today, this would be about $800, which is considerably less than the estimated current value of the island's land, which is in the ballpark of $1.7 trillion.)[7]

Minuit's worldview was informed by European treaties and laws, which included the ownership of real estate. His goal was to attract settlers to the colony of New Amsterdam, and outright possession of the land would help in this effort. The Lenape and other native Americans, however, are thought to have had fundamentally different ideas about the concept of ownership. For them, the land, like the air and the water, was available to everyone and could not be bought and sold. The Lenape didn't even live on the island; they only hunted and fished there. They probably regarded Minuit's trade goods as gifts of friendship or, at most, an agreement to allow the Dutch colonists to use the land.[8]

As these examples illustrate, implicit assumptions can derail communication despite the best of intentions. The Dutch and the Lenape didn't think it was necessary to question one another about their respective views on landownership rights. Minuit simply assumed that such things were transferable, while the Lenape culture lacked the concept on which the so-called purchase was negotiated.

However, we apparently fare no better when it comes to modern-day assumptions about how to measure units of force. The engineers at Lockheed Martin were a product of their industry's culture, and in that world, customary units prevailed. At NASA and JPL, metric measurements had always been employed. Even when deviations were noted in the trajectory of the Mars Climate Orbiter on its flight to Mars, no one seems to have suspected that mismatched units of measurement might be the culprit.

Frames of reference are useful because they allow communication to proceed efficiently. Conversation would grind to a halt if we constantly had to check the conceptual underpinnings of our partner's worldview. These frames, however, are insidious precisely because they are invisible. And we must occasionally pay the price for such assumptions when they turn out to be wrong.

GREAT EXPECTATIONS

On the evening of June 10, 2007, millions of Americans settled in to watch the series finale of *The Sopranos*. The HBO drama about the New Jersey mobster and his family had been incredibly popular since the first season had been broadcast eight and a half years earlier. It was expected that the finale would provide closure on a number of issues that had arisen during the series' 86-episode run. How would it all end?

During the final minutes of the last episode, viewers watched expectantly as one by one, the members of Tony's family arrive at a diner. Many fans were nonplussed, to say the least, when there was an abrupt change to a black screen without sound. More than a few viewers assumed that there was some sort of technical problem and that they were missing the final revelations about the fate of Tony and his family. However, after 10 seconds of black screen, the episode's credits began to roll. It was over, but what happened? Was Tony whacked in the final scene, as many began to speculate? The show's creator, David Chase, wouldn't say, and speculation has continued for years.[9] This episode and others like it (*Game of Thrones*, anyone?) make clear that people have strong expectations about what should and should not happen in a series finale and are disturbed when those expectations are violated.

The idea that expectations are an important part of human behavior has a long history in psychology. In the 1920s, for example, the British psychologist Frederic Bartlett used the term *schema* (from the Greek for "form" or "figure") to refer to mental structures that we use to make sense of our experiences. In a series of studies, he showed that these schemas exerted a powerful influence over how we interpret and remember information.

In his experiments, Bartlett had Cambridge students read a Native American story called "The War of the Ghosts." This folktale contains a number of elements that make little sense outside the cultural context in which it was handed down. In the story, a boy who is hunting seals is recruited by others to help them make war on a neighboring group. A battle is described, and some supernatural events take place. The story ends with the boy returning to his village, his recounting of the events of the battle, and his sudden death.

Bartlett asked his study participants to recall the story, as accurately as possible, either shortly after reading it or at varying lengths of time afterward. He found a characteristic pattern of distortions in these reproductions. For example, less familiar elements were changed to concepts that were more familiar to the participants: "hunting seals" became "fishing," while "canoes" turned into "boats." Bartlett's participants also frequently omitted the stranger and harder-to-understand elements, such as the ghosts that are mentioned during the battle or the fact that something black came out of the boy's mouth before he died.

To explain this pattern of results, Bartlett proposed that the expectations of his participants were responsible for these distortions and omissions. Specifically, their culturally based schemas caused them to convert the less familiar to the more familiar, and elements that didn't fit their schematic frameworks were discarded. In short, their expectations exerted a powerful effect on what they were able to remember about the story they had read.[10] Bartlett's work was rediscovered by cognitive scientists in the late 1960s and has exerted a great deal of influence on modern theories of human memory.[11]

A similar idea can be found in the work of computer scientists and artificial intelligence researchers, who developed the notion of *scripts*. These are event schemas, such as one's morning routine or the elements of having a meal in a restaurant.[12] We intuitively understand that the elements in such scripts have a temporal sequence: we must be seated before we order, and we pay for the meal after it is eaten. Humans acquire these scripts by repeating these activities many times, and the whole sequence becomes second nature—a restaurant script. An artificial intelligence agent, on the other hand, would need to have the event sequence explicitly coded into its programming since it would have no real-world experience to draw on.

Schemas and scripts may sound like abstractions without much real-world importance, although they might help you perform well on the television program *Family Feud*. On that game show, contestants are asked, in essence, to make their schematic knowledge explicit ("Name things you would find at the circus"). But outside of that limited domain, do they have any real utility? It's been argued that our schemas can blind us to events in the world, causing us to misinterpret what we see and hear. A particularly poignant example of this occurred right before the Japanese attack on Pearl Harbor—the event that plunged America into World War II.

On the morning of December 7, 1941, the U.S. military was operating a number of experimental radar installations on the Hawaiian Islands. One of these, the Opana Station on Oahu, was 15 miles north of Pearl Harbor and crewed by two army privates. At 7 a.m., they were completing a three-hour training exercise when they observed a large signal on their radar oscilloscope. It was so large that one of them, Joe Lockard, thought that the equipment must be malfunctioning. The other soldier, George Elliot, decided to contact the Aircraft Warning Information Center.

Lieutenant Kermit Tyler was the officer on duty, but he had been with the center for only two days and was not trained in using radar. He listened to Lockard's description of the unusual signal from an unusual direction. And Tyler's reply to Lockard? He told him, "Don't worry about it." Tyler's response was based on his incorrect belief that the radar had detected a group of B-17 bombers that were scheduled to arrive that morning from the U.S. mainland. As a result, he decided not to contact his supervisor. Less than an hour later, the Japanese aircraft that Lockard and Elliot had detected began to rain bombs on the U.S. fleet anchored at Pearl Harbor. The attack killed more than 2,400 Americans and sank or damaged 21 ships.[13]

We can see throughout this example how the expectations of those involved influenced their conclusions and decisions. An unusually large radar signal is initially dismissed by one of the soldiers as an equipment malfunction. The lieutenant decides that the privates' report is of no importance because an explanation is readily at hand: the radar must have detected the arrival of U.S. aircraft.

Psychologists have gone on to conduct many studies that underline the importance of assumptions as drivers of behavior. Expectations about events are incredibly helpful, but as we have seen, our ideas about what *should* be the case can be a source of misunderstanding as well.

THE VAGARIES OF VAGUENESS

The impeachment trial was heated and contentious. Members of both political parties charged one another with sanctimonious behavior, acting in bad faith, and engaging in political grandstanding. In the end, however, the House of Representatives, voting largely along party lines, found that the president of the United States was guilty of serious crimes. Following an equally acrimonious trial in the Senate, however, the nation's chief executive was ultimately acquitted.

This exact scenario has played itself out four times in American history, most recently in 2021, when Donald Trump was accused of inciting an insurrection. Two years earlier, he had been charged with abuse of power and obstructing Congress. In 1998, the charges against Bill Clinton were the same, although he was also accused of perjury. But in the nation's first impeachment trial, held in 1868, the charges against Andrew Johnson revolved around a decidedly different issue: had the president violated the Tenure of Office Act?

Simply put, this law prohibited the president from removing members of his administration when the Senate was not in session. It had been passed by the House and the Senate over President Johnson's veto. A few months later, while the Senate was adjourned, Johnson removed Secretary of War Edwin Stanton and temporarily replaced him with Ulysses Grant. A major issue during the impeachment trial was whether Johnson had truly violated the act. Stanton was a holdover from the Lincoln administration, and it was not clear whether the law applied to him. In the Senate trial that followed, Johnson managed to avoid removal from office by a single vote.

In 1869, after Grant became president, the Tenure of Office Act's requirements were weakened by Congress. It would go on, however, to create problems for Grover Cleveland in 1886, and Congress ended up repealing the law in its entirety the following year. The original rationale for the measure had been to check the power of the chief executive, but in the end, it was seen as being too imprecise to be enforceable.[14]

The debate about how to interpret the Tenure of Office Act exemplifies how vagueness allows for different interpretations of the same statement. This has been a perennial issue in debates about laws and

statutes. Vagueness has, for example, been at the crux of drafting legislation that demarcates the line between child pornography and innocent
family snapshots. Does a naked child squatting on a potty constitute an
instance of "lascivious exhibition"—a term used in statutes to define ille-
gal images?[15] How about a painting of a winged cherub? And is nudity a
requirement? The U.S. Third Circuit Court of Appeals didn't think so
in its ruling on this subject.[16]

To address such conundrums, courts have made use of a "void-for-
vagueness" doctrine to strike down a number of poorly written statutes.[17]
For example, a Florida ordinance that was intended to allow police to
arrest vagrants had to take on the formidable task of defining the concept
of vagrancy. Among other suspicious behaviors, the statute included
the act of "wandering or strolling around from place to place without
any lawful purpose or object."[18] In practice, this could include virtually
anyone, from desperate fugitives to innocents seeking a little fresh air and
exercise. And in a 1972 case, *Papachristou v. City of Jacksonville*, the
Supreme Court ruled that this description was unconstitutionally vague.
In a unanimous ruling, the court declared that the statute granted law
enforcement an almost unlimited power to make arrests.[19]

The U.S. Constitution itself is replete with vaguely worded statements
and underspecified concepts. It's worth noting that the entire document,
including the original 10 amendments, runs to fewer than 5,000 words.
(The 17 additional amendments ratified to date constitute an additional
2,600 words.) That's not a lot of verbiage to describe the workings of a
federal government, not to mention enumerating the rights of the states
and the citizenry. However, it would have been impossible for its authors
to anticipate every possible future scenario. Instead, the framers viewed
their handiwork as scaffolding that could be elaborated on by legislation
in Congress, interpretation by the courts, and further amendment. James
Madison, in *Federalist Paper No. 37*, explicitly stated that all new laws
are "obscure and equivocal" and that a process of adjudication is necessary to work out their meaning in practice.[20]

Nonetheless, some U.S. Supreme Court justices, such as Clarence
Thomas and Neil Gorsuch, style themselves as originalists: they believe
that the Constitution should be interpreted strictly as intended by its
authors.[21] This involves a certain degree of hubris because there is no

way of knowing whether a justice has correctly divined the intentions of individuals living several centuries earlier. Ideas about originalism, combined with the brevity of the constitutional text, have resulted in vigorous debate about the framers' intent on issues like the right to keep and bear arms—the 27 problematic words that make up the Second Amendment.

Some legal scholars have inveighed against the use of dense, lengthy boilerplates found in many legal contracts and settlements.[22] The poster child for such screeds is the end-user license agreement that accompanies software. Their length and opacity deter all but the most determined to make sense of them.[23] One purpose of such agreements is to shield the developers from liability. A primary reason for their length, however, may be to avoid the sorts of vague and ill-defined language that doomed the Tenure of Office Act and almost unseated a president.

HINTS AND DOG WHISTLES

The greatest thing in family life is to take a hint when a hint is intended—and not to take a hint when a hint isn't intended.

—Robert Frost (attributed)[24]

In some cases, misunderstanding occurs because speakers choose, for a variety of reasons, to deliberately shroud their statements in ambiguity. And although there might be a good reason for such vagueness, their recipients might not always get the point.

Let's start by considering the case of hints. If we're asking someone to do something for us, we probably want to avoid being too blunt in making the request. After all, no one likes being ordered around or told what to do. Different cultures, however, differ in the degree to which making direct requests is considered acceptable. In English-speaking countries, a politeness norm dictates that such requests be expressed indirectly.[25] This occurs so frequently that we might not even realize how oblique such statements truly are.

Imagine, for example, that two people are sharing a meal at a restaurant. While dining, one of them might ask the other, "Can you pass the

9

salt?" Although it's clear what the speaker wants, the request is literally a question about ability, not a direct command to do anything. But it's a polite way of making such an appeal. In a similar way, queries like "Do you know what time it is?" or "Would you mind taking our picture?" can be interpreted literally as questions that require a yes or no reply. Nevertheless, speakers of English immediately understand these so-called indirect speech acts for what they are: courteous ways of asking for things, actions, or information. Nonnative speakers of English, on the other hand, may need to be explicitly taught about this expectation for indirectness.[26]

In other cases, a statement or a question may have multiple interpretations. An utterance like "It's getting really hot in here" can be an indirect request to have someone adjust the thermostat or open a window. However, it could also be a sarcastic observation about a cold room. Under the right circumstances, it could even be a flirtatious come-on as the speaker suggestively loosens an article of clothing. And a statement like "I've washed the dishes" might serve as an indirect way of telling a spouse to take out the garbage.

Hints are also employed to make clear a lack of interest in the attentions of others. If someone asks you out on a date, you might reply, "Sorry, I have to study" (or even "I have to wash my hair," a mid-twentieth-century go-to line). Such circumlocutions are meant to let the other person down in a face-saving way. A socially maladroit suitor, however, might fail to take the hint. In such situations, a less ambiguous response, like "I have a boyfriend," might be required to ward off a more persistent admirer.

Ambiguity can be useful in other contexts as well. A political candidate might choose to make statements that implicitly invoke certain stereotypes. This is common when the topic involves politically sensitive issues. For example, during the 1976 presidential campaign, candidate Ronald Reagan frequently mentioned the case of Linda Taylor, a woman who the *Chicago Tribune* referred to as a "welfare queen" for abusing public assistance programs. Reagan would refer to her as "that woman in Chicago." His supporters could interpret this as a critique of minorities and poor people who were on the dole.[27]

Why do politicians use such indirect speech? A primary reason is that it gives them plausible deniability. Reagan, for example, could truthfully assert that his appeals for welfare reform weren't racially motivated because he never referred to race. Reagan's supporters, however, would hear the message that only they were attuned to hear: a coded reference to the idea that minorities abuse public assistance. For that reason, such appeals have come to be called "dog whistle politics." Another reason that politicians employ them is that they've been shown to work—such coded racial messages have even been shown to affect the opinions of people who are politically liberal.[28]

Throughout U.S. history, a number of phrases have been employed as dog whistles for politically incorrect positions. References to "states' rights," for example, have been used to signal implicit support of white supremacy. During his presidential campaigns, George Wallace made use of the phrases "law and order" and "neighborhood schools" as codes for racial segregation.[29] In the 1980s and 1990s, candidates touting "family values" were narrowcasting anti-LGBTQ attitudes or support of anti-abortion legislation to the fundamentalist community. Although no one opposes states' rights, law and order, or family values in principle, these coded messages were understood clearly by their intended audiences.

In some cases, a dog whistle can be so obscure that it is incomprehensible to the general public. President George W. Bush was well known for using coded language that appealed to his conservative Christian base. Phrases like "wonder-working power," which appeared in his 2003 State of the Union address, were intended to catch the attention of those who were familiar with the evangelical Protestant hymn "There Is Power in the Blood."[30]

In other cases, there may be some dispute as to whether a word or phrase is truly intended as a dog whistle. Consider the case of Josh Hawley, a Republican senator from Missouri, who in July 2019 made repeated references to "cosmopolitan elites" in a speech he gave at a meeting of conservatives in Washington, D.C. Among other assertions, he claimed that this cosmopolitan class was weakening the United States through its control of big business and its "international network." A number of commentators criticized the speech as anti-Semitic, pointing out that both Hitler and Stalin had used the term "cosmopolitan" as a slur against Jews.[31] Some Jewish leaders called Hawley out for using the term, and he responded by tweeting that "the liberal language police have lost their minds." The speech also had its defenders, such as those who argued that "cosmopolitan" was meant in its more general, citizen-of-the-world sense and was not intended to be anti-Semitic.[32]

As the philosopher Ian Olasov has pointed out, many expressions function as coded messages because of the positive or negative emotions they engender. Even an ostensibly neutral phrase like "women and children" might be used strategically to invoke the blameless victims of some tragedy.[33]

HOW GOOD IS GOOD ENOUGH?

The perfect is the enemy of the good.

—Voltaire, *Dictionnaire Philosophique* (1770)

Close enough for government work.

—Twentieth-century American idiom

When is a stone small enough to be called a pebble? When is it large enough to be called a boulder? How much hair must a man lose before he qualifies as bald? There are no objective answers to questions like these, and reasonable people may disagree about the precise thresholds that are required. But such disputes highlight another potential cause of miscommunication: the words that we have at our disposal are not always specific, precise, or well defined.

If we seek enlightenment by turning to a dictionary, we are likely to come away disappointed. In the case of "pebble," for example, we might find it defined as a small stone. The word "small" is problematic because it is a comparative rather than an absolute judgment: smaller than what? And if being bald is defined as having lost most of one's hair, the meaning of "most" is equally problematic.

In theory, the problem of imprecision could be solved by bulking up a given language with hundreds or even thousands of new words to cover the gaps that currently exist between concepts. A "wuzzle," for example, could denote a rock that is bigger than a baseball but smaller than a basketball. "Chabber" could mean having lost all of one's hair except for a horseshoe ring around the top of one's head.

But would this truly solve the problem of linguistic precision? Dictionaries are already larded with thousands of words and phrases that people don't use, either because they don't know them or because the distinctions that they make aren't particularly useful. (For example, the phrase "Hippocratic wreath" already exists to refer to the horseshoe ring form of baldness mentioned earlier, but few people use it.)

In other cases, the issue may not be the precision of someone's language but rather how much of it they choose to employ. The cognitive scientist Herb Simon proposed that people do things that are good enough for the purpose at hand. He coined a useful term for this as well, proposing that people engage in "satisficing."[34] Although Simon was referring to the domain of decision making, this idea can be applied to communication as well. We see a similar notion in the writings of the philosopher Paul Grice.[35] In his take on the ground rules for conversation, he described a "maxim of quantity": don't say more than is required. In many cases, we simply don't need to convey a great deal of information to communicate successfully.

We see such ideas playing themselves out in many of our social interactions. If you need to explain to your partner why you are late in meeting her for coffee, telling her that the line at the supermarket was long is good enough. You could, in theory, specify the minutes you had to wait in line, the number of shoppers who were in front of you, and the exact number of items that the cashier had to scan before ringing up your purchases. But this would be overkill for the purpose at hand: providing a credible reason for why you were running late. (And in this case, an overly detailed excuse might seem suspicious.) We satisfice, and then we get on with our lives.

A countervailing narrative, however, has taken shape with regard to Grice's ideas about quantity. Saying only what is needed for current purposes may be efficient, but it appears that, in some cases, speakers choose to be more verbose than is strictly necessary. Paul Engelhardt and his collaborators, for example, found that research participants tended to overdescribe objects when they were asked to provide directions to another participant. They said things like "Put the apple on the towel in the box" even though there was only one apple in front of them, making the phrase "on the towel" redundant.[36] The philosopher Paula

Rubio-Fernandez has obtained similar results and suggests that a variety of factors, such as pertinence or predictability, might influence whether people use more words to describe something than is strictly necessary.[37]

Choosing how much to say in a given situation can also be thought of as a discourse style, and there may well be a variety of contextual, social, and even personality factors at work. Jean-Marc Dewaele has found that when a situation is perceived as being relatively formal (as in taking an oral exam as opposed to chatting about hobbies), people tend to become more expansive in their discourse style. Female participants in his study were more explicit in the informal situation, while extraverted participants were more verbose in the formal situation.[38]

Deciding how much information to communicate may vary by topic as well. As early as 2006, Warren St. John was lamenting in the *New York Times* that younger generations were divulging too much information (TMI) about their personal lives on social media.[39] But scientists also have to grapple with TMI when they attempt to explain complex issues like climate change to the public. Kevin Finneran has pointed out that full disclosure can easily give way to information overload, and therefore a delicate balancing act is required.[40]

There may be a host of situational factors that affect whether Grice's maxim of quantity will be followed. In some situations, a more detailed accounting might be expected and might even be absolutely essential. If you had been running late to get your very pregnant partner to the hospital instead of meeting her for coffee, a more detailed explanation for your tardiness would probably be called for. And if you were following instructions for defusing a ticking time bomb (as one does), you might be disconcerted by a command like "Cut the blue wire" if several of the wires in the device were blue. In a critical situation, you would expect extremely specific guidance, such as "Cut the pale blue wire below the green wire that's attached to the left side of the timer." In short, what might seem like overkill in one situation might be underkill in another.

So far, we have seen how frames of reference, expectations, and vagueness can lead to communication failures. Let's conclude our initial foray into the causes of miscommunication by considering how implicit assumptions held by people in different groups affect their electronic communications.

WHEN INFORMAL IS TOO INFORMAL

During the early years of the internet, a set of customs and conventions—referred to as "netiquette"—evolved among chat room denizens and those sending electronic messages. Compared to the current social media landscape, this earlier online world was a decorous and relatively genteel place. The rise of instant messaging and then texting via cell phone, however, served to erode the distinction between more formal modes of online communication and more relaxed and casual messaging.

As time went by, a digital divide became apparent. Individuals who were already adults in the mid-1990s, when the web transformed the cultural landscape, had different expectations about communicative norms than those who were born after the early 1980s.

For this latter group, sometimes referred to as digital natives, the internet has always been part of their lives, and their familiarity with this world and its customs has been much less influenced by pre-internet communication conventions than it was for their elders, the so-called digital immigrants. The immigrants' expectations are often violated by the natives, and the natives themselves have developed new conventions that are not recognized or even apparent to older generations.

As with any broad generalization, it is problematic to assign the term "digital native" to everyone born after, say, 1980.[41] An even more problematic assumption about digital natives is that such individuals process information in fundamentally different ways or that they are more adept at multitasking.[42] A more fine-grained analysis can be found in the work of internet linguist Gretchen McCulloch, who has identified five distinct cohorts based on an online survey that she carried out in 2017.[43]

Old Internet People, as McCulloch refers to them, are the pioneers: those who first ventured online in the 1960s, 1970s, and 1980s. They interacted with other colonists via Bulletin Board Systems (BBSes), Multi-User Dungeons (MUDs), and the Usenet through slow dial-up modem connections, often connecting via service providers like CompuServe or Prodigy. As a group, they are technically savvy (they had to be), and many can write computer programs (they can code apps, to use modern parlance).

A second group is Full Internet People, who, by contrast, came online as the web exploded in popularity in the mid-1990s and replaced the older internet. For them, the social aspects of the online world were paramount, and they made extensive use of various messaging platforms, such as AOL Instant Messenger (AIM), and early social networking sites, like MySpace, although they were still getting online mostly via service providers like America Online (AOL). Later, they moved on to Facebook and Twitter.

Semi Internet People, a third cohort, went online with the emergence of the web, as did Full Internet People, but for them, it doesn't exist at the center of their social lives. They often use the internet for professional purposes or because their workplace requires it. They are more likely to interact with others via email than to follow them on Facebook.

A fourth group can be thought of as Pre-Internet People. They are typically older and got online later, after the Full and Semi Internet cohorts. They may venture online only for a single purpose, such as to access Facebook, and they aren't particularly familiar with internet slang or other linguistic conventions.

Finally, there are the Post Internet People. They represent the youngest group and consist of the ever-expanding cohort that grew up in an ever-present, always-online world. Their social relationships are heavily dependent on texting and (as of this writing) Snapchat and Instagram. Instead of migrating to the web from a desktop or laptop computer, they may have always interacted with the online world via a smartphone.

McCulloch argues that a key distinction among these cohorts is whether their informal writing is influenced more by offline norms, which they may have acquired via formal education, or the online norms commonly employed in social media. When members of a given cohort communicate with one another, they make use of shared communicative norms. Problems can arise when members of one cohort interact with members of another, and this can be thought of, yet again, as a strike against the chances for communicative success. This makes communication brittle and liable to break if there are additional difficulties.

One arena in which this has been studied is how university students interact with their instructors via email. A 2009 study found considerable irritation on the part of the faculty caused by their students' "overly casual" linguistic choices. The researchers found that instructors were particularly bothered when students failed to sign their messages and when they used texting conventions, such as "RU" for "are you."[44]

On the other hand, Post Internet People make use of texting conventions that may be unknown to Old, Semi, or Pre Internet People. To cite but one example, it has been demonstrated that text messages ending with a period are perceived as less sincere than those that do not.[45] And short ambiguous responses, such as "maybe," are perceived as more negative when they are followed by a period.[46]

Even the most common of all abbreviations—for "okay"—has undergone a transformation in the world of texting. It is now perceived as overly formal, and alternatives are commonly employed instead. One form, "kk," is neutral and employed to simply indicate that a message has been received.[47] "Okay" or (especially) "k," however, is often perceived as an aggressive or hostile form of acknowledgment. As one Twitter user put it, "I get that my parents don't understand what texting the letter 'K' means but it still hurts."[48]

Many Old or Semi Internet People still carefully capitalize words, use apostrophes, and employ terminal punctuation, much as if they were writing a more formal missive. They may not, however, be aware of the impression they are creating by writing in this way. For them, the terminal period is a dog whistle that they themselves cannot hear, even as it is perceived clearly and pejoratively by their Full and Post Internet addressees.

2

PSYCHOLOGICAL FACTORS

Let's now dig a little deeper into some of the psychological factors that lead to miscommunication. As we will see, no one is immune to such effects. Alexander Pope may have written that "a little learning is a dangerous thing," but the reverse is also true: when we bring a lot of prior experience or information to an interaction, we may experience difficulties because of the so-called curse of knowledge. In general, we experience fewer problems in our communication with family members and friends because we have a better understanding of what they do and do not know. When communicating with strangers, on the other hand, all bets are off, and these issues are magnified when we resort to electronic communication, such as email or social media.

DO YOU SEE WHAT I MEAN?

Imagine that you are a participant in a psychology experiment. The researcher asks you to pick a well-known song from a list, such as "My Country 'Tis of Thee" or "Silent Night," and then to tap out its rhythm. In addition, you are asked to guess the likelihood that another person would be able to figure out the identity of the song you chose based only on your tapping performance.

When a group of Stanford undergraduates answered this question, their average estimate was 51 percent. However, when another group of participants served as listeners and tried to name the songs based solely on these rhythmic taps, their performance was abysmal. Only three of 120 tapping performances were identified correctly, yielding a success rate of 2.5 percent. In other words, the participants who had tapped out the rhythms had overestimated their own prowess by a factor of 20.[1]

The results from this study dramatically illustrate a cognitive bias known as the curse of knowledge. In general, we assume that others possess the requisite information to understand things we already know. Research participants who tap along to their mental rendition of "Silent Night," for example, clearly find it difficult to imagine others not being able to recognize the familiar melody, even though identifying a song from its rhythm alone is extremely difficult. But that's the trap—once we know something, it becomes surprisingly difficult for us to imagine *not* knowing it. As William White put it, "The great enemy of communication, we find, is the illusion of it."[2]

Working hand in hand with the curse of knowledge is another cognitive distortion known as the egocentric bias. This refers to a general tendency for people to rely too heavily on their own perspective or point of view.[3] In general, this bias may be adaptive, but it also interferes with our ability to see the world through someone else's eyes. And it leads to other predictable biases, such as a pervasive belief that our own opinions and values are more popular than they truly are: the so-called false consensus effect.[4]

One consequence of these distortions—the curse of knowledge and the egocentric bias—is overconfidence. As we saw with the tapping Stanford students, people tend to overestimate the degree to which their own knowledge exists in the minds of others. "If I can hear 'Silent Night' clearly in my head," they seem to think, "then I'm certain someone else could identify it based on my rendition of its rhythm." However, as we have seen, others cannot.

Another consequence of these cognitive biases is the illusion of transparency. That is, we believe that our internal states are more apparent to others than they are in reality. An actor, for example, may worry excessively that her nervousness will be perceptible to members of an audience. For this reason, this is sometimes referred to as the spotlight effect.[5] As the authors of one research study put it, it's as if we believe that our thoughts and feelings "leak out" of us and are more visible to others than they truly are.[6]

You might suspect that children and adolescents would be more prone to such cognitive biases than adults are. In comparison to their elders, teenagers have had less experience with the social world and therefore have had fewer opportunities to see how different that world may appear to others. However, one study of the illusion of transparency found that adolescents were about as egocentric as adults: both groups believed that lies they told could be detected more easily than was actually the case.[7]

At the same time, however, we overestimate how well we can identify another person's mental and emotional states—a phenomenon referred to as the illusion of asymmetric insight. This distortion manifests itself when we think we know our friends and roommates better than we think they know us. This illusion also exerts its effects at the group level: people believe that the faction they belong to understands other groups better than members of those groups understand them.[8]

Not surprisingly, these cognitive biases and their consequences have important implications for how we communicate and for why miscommunication occurs. When we talk to other people, we must make all sorts of assumptions about what they know and don't know, and it's easy to get this wrong. When we assume that little knowledge is shared, we may end up providing lengthy explanations that serve only to frustrate and annoy our conversational partners. For example, it can be tedious or even exasperating when a coworker describes a complex procedure that is already familiar to us.

An even greater conversational failing, however, occurs at the other extreme when we assume that others know more than they do. And research by psychologists suggests that people routinely overestimate their communicative effectiveness.[9] The cognitive biases reviewed in this section help to make clear why we are overconfident and frequently make unwarranted assumptions.

Monitoring the amount of knowledge that we share with others is an important topic that deserves to be discussed on its own. Therefore, we will turn to this subject in the next section. We will also return to a discussion of egocentric biases in chapter 10 when we consider why miscommunication frequently occurs when email, texts, and social media posts are employed. As we will see, these cognitive biases, combined with a lack of nonverbal cues and feedback from the recipients of our electronic messages, can create the conditions for a perfect storm of miscommunication.

CREATING COMMON GROUND

As we will see throughout this book, communication failures are frequently the result of mistaken assumptions about common ground. Cognitive scientists define common ground as the knowledge, attitudes, and beliefs that two people share. This information regulates how they speak to one another in a number of important ways.

Let's consider a concrete example. Imagine that you're talking to a new coworker for the first time. (To help keep things straight, let's call her Alice.) You might tell Alice about your family, where you live, or how you spend your free time. However, you would probably be careful not to make unwarranted assumptions that could end up creating confusion. You might even find yourself repeating certain key bits of information. For example, you might make reference to your husband and tell Alice that his name is Bill. And later in the conversation, you might refer to him again by specifying both his name and his relationship to you.

If you were to encounter Alice on the following day, how would you refer to your spouse? Chances are that you would simply say "Bill" because his relationship to you should now exist as part of your shared common ground with Alice. And this should be true for other important information that you've told her. But imagine that you mentioned your husband Bill only once and then didn't get a chance to talk with Alice for several weeks. In that case, you might well find yourself falling back to referring to him as "my husband Bill" because you're uncertain whether Alice will remember who he is.

Throughout time and through additional conversations, you and Alice will develop an extensive body of shared knowledge and experiences. This might include the names of family members, likes and dislikes, hopes for the future, and inside jokes. Even if you don't become close friends, these interactions will fundamentally change the ways that you converse with one another. When people share a lot of common ground, their referents tend to be concise and elliptical. They can become so insular that a stranger eavesdropping on such a discussion might have a hard time understanding it.[10]

To more fully understand the concept of common ground, let's consider what "shared" means in this context. Imagine that, in your first confab with Alice, you don't refer to your husband at all. And before you speak with her again, she has a conversation with your boss, who happens to mention that you are married to someone named Bill. Technically, this information is now shared in the sense that both you and Alice know that a guy named Bill is your spouse.

But if you don't know that Alice knows this, then it's *not* part of your common ground. As a result, in referring to Bill, you would helpfully add "my husband" to clarify who he is. And if Alice were to reply, "Yes, your boss mentioned that Bill is your husband," then—and only then—would Bill's name and relationship to you become part of the shared common ground. You know it, Alice knows it, and, crucially, you *know* that Alice knows it. This process by which common ground is built is called grounding.[11]

In general, people are good at keeping track of shared common ground. That is, they usually provide some context, like "my husband," when new information is referred to. In addition, they don't keep repeating it when shared information is well known to both parties. Imagine how annoying it would be if your sister referred to "our brother, John" every time she made reference to him. You might find this a bit maddening unless you also happen to have a cousin and a mutual friend who all share the name John.

Psychologist Herb Clark, who has written extensively on this topic, has described different types of common ground.[12] For example, things that are perceptually salient in the physical environment automatically become part of the common ground: you and your conversational partner share them because both of you are experiencing them. Imagine standing next to a stranger on a sweltering subway platform. You could try striking up a conversation by alluding to the extreme temperature, such as by saying, "Hot enough for you?" (In this case, the other person would certainly know what you are talking about. Whether or not such a conversational gambit would be welcomed is another matter entirely.)

Common ground can also be inferred when membership in certain groups has been mutually established. If two strangers discover that both are dentists, for example, they should feel free to refer to technical terms or procedures that are common to that profession, even if these are not well known by civilians.

And while shared occupations can function in this way, the same holds true for many other groups and communities. Membership in religious sects and alumni organizations or even living in a particular neighborhood are all examples of this. Even certain kinds of experiences, like being a plane crash survivor, a mother, or a drug addict—not that these things go together—might qualify. Many getting-to-know-you conversations involve a search for such commonalities that can establish common ground.

Issues of common ground figure prominently in miscommunication because it can be hard to get the balance right. On the one hand, assuming common ground when it doesn't exist can create ambiguity and confusion. To return to our earlier example, a reference to someone named Bill with no other identifying information can be problematic. And when someone does this frequently, we might conclude that they are self-absorbed or egotistical.

On the other hand, when someone continually explains to us that John is their uncle and an airline pilot, we may experience some hurt feelings. Don't they remember that they told us about Uncle John the pilot yesterday and a couple of times last week? Do they think so little of us that they can't remember what they've discussed with us?

Added to this complexity is the fact that the world is a constantly changing place. People get married, but they also get divorced. As a result, tagging "Bill" as someone's husband is not necessarily true for all time.[13] And if two conversational partners know lots of people named Bill, then some sort of disambiguation may always be required.

And these problems only multiply if we consider other forms of communication. Authors, for example, must make educated guesses about what their readers know and do not know. Teachers must laboriously build up common ground between themselves and their students. In short, even under the best of circumstances, the process of constructing common ground is fraught, and failures in grounding will loom large in many of the examples described in later chapters.

JUST BEING SARCASTIC

Customer tweet: You are doing GREAT! Who could predict heavy travel between #Thanksgiving and #NewYears Eve. And bad cold weather in Dec! Crazy!
Airline response: We #love the kind words! Thanks so much.
Customer response: Wow, just wow, I guess I should have #sarcasm

—Exchange of tweets between a customer and an
airline service representative[14]

Perhaps no form of language is more prone to miscommunication and misunderstanding than sarcasm.[15] By saying the opposite of what they truly mean, speakers create a dilemma for their audience. Should they accept what they're hearing at face value, or should they go beyond the speaker's words and try to infer some other meaning?

We will see similar issues with regard to double entendre and euphemism, topics that will be discussed in chapter 5. For all of these types, it is the nonliteral nature of such statements that make them particularly susceptible to misinterpretation. In addition, as we saw in the first section of this chapter, people have an egocentric bias. This causes them to believe that they communicate better than they do in reality. In the case of sarcasm, the psychologist Jean Fox Tree and her collaborators had documented that, in many cases, such remarks are simply not perceived as sarcasm by others. The researchers refer to this gulf between the speaker's intention and the listener's incorrect interpretation as a sarchasm.[16]

Another aspect of egocentrism is the assumption that other people will communicate as they themselves do. The study conducted by Fox Tree found that the participants in her study expected their conversational partners to be about as sarcastic as they themselves were. Previous research has shown that while some people use sarcasm frequently, others employ it far less often.[17] Therefore, a mismatch in sarcasm use is likely to be particularly problematic.

We can assume, therefore, that literal-minded people will fail to detect much of the sarcasm that they encounter. At the same time, people with a predilection for nonliteral language may mistakenly perceive sarcasm when none was intended. The likelihood of miscommunication in such cases is high.

When speakers express themselves sarcastically in face-to-face inter-actions, they can make use of certain facial expressions or a specific tone of voice to signal what they mean. For example, rolling one's eyes or speaking slowly and loudly can function as behavioral cues for sarcasm.[18] Face-to-face interactions are also likely to occur between people who have at least some shared history together. As a result, listeners can rely on shared common ground to help determine whether their conversa-tional partner is being sarcastic.

But when an interaction occurs online, such as through social media, the communicative context is greatly impoverished. In such cases, shared common ground is likely to be low or completely absent. Behavioral cues are missing as well. As a result, the communicative process already has one strike against it. There are, however, a number of text-specific surrogates that can be pressed into service to replace the missing facial and vocal cues.

The tweet at the beginning of this section contains several instances of this. Typing in all caps can serve as a stand-in for speaking loudly. Exclamation points can function as a proxy for heavy verbal emphasis. Rhetorical questions ("Who could predict"), a tactic common in face-to-face sarcasm,[19] can also be exploited. Sarcastic statements in the offline world tend to make use of hyperbole, and extreme statements are com-mon in such language ("crazy") as well.[20]

But as we see in the response to the customer's tweet, all of these cues proved to be insufficient. It's possible that the comment from the airline was generated by a chatbot as opposed to a human being. Chatbots are automated artificial intelligence programs used by large corporations to create engagement with customers online. With regard to picking up on subtlety and nuance, however, they remain a work in progress.[21]

A chatbot makes use of algorithms to assess the sentiment of a state-ment and then attempts to respond appropriately. It certainly is true that "You are doing great!" could have been intended as a sincere statement. It's unlikely, however, that any human being who read the tweet in its entirety would have concluded that it was complimentary. Humans know that travel at the end of the year is typically extremely heavy. They also know that December is likely to be cold. Chatbots can "know" those things as well, but this sort of world knowledge must be added manually, which is a laborious process.[22]

To minimize miscommunication online, entirely new cues for signaling nonserious intent have been developed. Scott Fahlman of Carnegie Mellon University is credited with being the originator of the "smiley" in 1982. This is the sideways grinning face composed of the punctuation string :-), which functions as a marker for nonserious statements.[23] Collectively, these sideways faces are called emoticons, and emoji are their modern descendants.

Just as an emoji can help disambiguate nonliteral statements in email and texting, hashtags have evolved to serve this function in social media. A hashtag can refer to almost anything, such as a person or an event. It can also be used to tag a statement that is intended nonliterally. Hashtags like #sarcasm, #irony, and #joking are commonly employed for this purpose. In other cases, faux markup language is used, such as /s, which is a common sarcasm marker on Reddit.[24]

The development of emoticons, emoji, and hashtags can be thought of as organic responses to the problems that are inherent to communicating online. They are a hopeful sign that language can evolve in ways that serve to minimize miscommunication and misunderstanding, but as we will see in chapter 6, we have not yet reached that promised land.

"WHAT IS A 'DANGEROUS THING'?"

On Valentine's Day 2011, fans of the TV program *Jeopardy!* were treated to a match that was unlike anything in the game show's history. Facing off on that episode were two of *Jeopardy!*'s greatest champions, Brad Rutter and Ken Jennings. And appearing between them onstage was a third competitor—a hulking rectangular box with a video screen for a face. This was Watson, the creation of a team of researchers led by engineers at IBM. As it turned out, the mini-monolith on the show's stage was only a stand-in: the room-sized bank of servers that powered Watson was located elsewhere.[25]

The goal of Watson's designers had been to construct an artificial intelligence program that could defeat human *Jeopardy!* opponents, and for everyone involved, there was a lot at stake. For the flesh-and-blood competitors, the winner would get bragging rights for all of humanity as well as a cool million dollars. If Watson won, IBM could lay claim to an entirely new level of artificial intelligence ability. And victory in the *Jeopardy!* competition would easily surpass the company's previous signature accomplishment—developing a chess-playing computer that defeated Gary Kasparov, the best human player, in 1997.

In comparison to its carbon-based competitors, Watson did have some important advantages. For one thing, it already "knew" all the answers. After all, Watson had access to about 200 million pages of information downloaded from the internet, including the full text of Wikipedia and the archive of the *New York Times*. It could also search this vast reservoir of knowledge at the blinding speed of a million books each second. What occasionally tripped Watson up, however, was its inability to truly understand the game's clues.

In some cases, Watson failed to pick up on the puns, double meanings, and other subtle wordplay that *Jeopardy!*'s clue writers frequently employ. For example, during a practice game, Watson responded with "What is milk?" to the clue "This trusted friend was the first non-dairy powdered creamer." (In case you're playing along, the correct question is "What is Coffee-mate?")[26] In another case, Watson responded to a clue about art with "Rembrandt" instead of "Pollack" because the answer specified a particular decade—the '40s—but not the century.[27]

During the two-game match, other cracks in Watson's façade appeared. For example, during the first game, one of the answers was "It was the anatomical oddity of U.S. gymnast George Eyser, who won a gold medal on the parallel bars in 1904." Ken Jennings rang in first and asked, "What is a missing arm?" but his response was incorrect. Watson then rang in with "What is a leg?" The host, Alex Trebek, ruled that this was also wrong since Watson didn't specify that the athlete was *missing* a leg. Watson didn't have the advantage of hearing Jennings's incorrect response—it lacked the ability to perceive speech—and appears to have been tripped up by the notion of an "anatomical oddity." David Ferrucci, the team leader for the Watson project, suggested that "the computer wouldn't know that a missing leg is odder than anything else."[28] Despite such errors, Watson went on to win the match, and its creators donated the computer's winnings to charity.

Watson's missteps, however, are instructive for highlighting yet another reason why misunderstandings happen. Although people don't have access to as much information as computers, they do possess a great deal of world knowledge that can make up for this—such as understanding that it would be unusual for an Olympic athlete to be missing a leg. And a person would also be more likely than a computer to understand that a "trusted friend" is a clue pointing toward words like "partner" or "mate."

People possess all sorts of implicit knowledge that they pick up through interactions with others and simply by living in the world. We learn that going to a coffee shop, for example, is generally regarded as a pleasurable activity; going to the dentist, however, is not.[29] We aren't taught such things but instead acquire this information through our experiences.

Computers can make use of such information, but to do so, implicit knowledge must be made explicit and then tediously coded into a machine's set of instructions. Artificial intelligence researcher Doug Lenat is creating a computer program in this way, and Cyc, as it is called, now consists of millions of such rules. However, this has been the work of decades, and there is still no end in sight.[30] Score one for the humans!

Psychologists who study language comprehension, however, have discovered that all of this world knowledge also has a downside: it can blind us to detecting errors in things we read and hear. And our world knowledge creates semantic "illusions" that can be as powerful as the perceptual kind.

The best-known example of this is called the Moses Illusion. When asked, "How many animals of each kind did Moses take on the ark?" most people will confidently answer "two," even though it was Noah and not Moses who sailed into that perfect storm in the book of Genesis.[31] Once the relevant biblical story has been activated in long-term memory, however, we fail to notice even blatant contradictions, such as the wrong name of the featured character.

You might expect that experts in a given domain would be less susceptible to the Moses Illusion than people without such knowledge. Experts can, after all, do things more quickly and easily than nonexperts,[32] which means that they should have the extra mental resources that are needed to spot errors or contradictions. In an experiment designed to test this hypothesis, however, experts still sometimes failed to notice semantic illusions in their areas of expertise. The effect was attenuated but not eliminated.[33] In short, a little knowledge is a dangerous thing—but so is a lot of it.

GUILT BY ASSOCIATION

In December 2019, the Manilla-based boy band SB19 began a 10-city tour through the Philippines. The previous month, they had become the first Filipino group to chart on *Billboard*, and Josh, Sejun, Stell, Ken, and Justin were excited about starting their tour. The first stop was Negros Occidental, a province of Western Visayas known as the "sugar bowl of the Philippines" because of its intensive sugarcane cultivation.

In a tweet posted on December 21, the band announced, "Hello, Negros! We are now in your zone!" Undoubtedly, this message was received with excitement by their fans. However, it was also met by incomprehension, derision, and accusations of racism as their greeting was retweeted thousands of times.[34] Even though Negros is the fourth-largest island in the Philippine archipelago, with a population of 4.4 million, most people living outside Southwest Asia had probably never heard of it. The island had been named by Spanish colonizers in 1565 in reference to the relatively dark-skinned inhabitants they encountered there.

Clearly, however, the band members were referring to the place and not making use of a dated and offensive racial term. These considerations, however, were lost on many members of the online community, some of whom reacted with outrage.

A few weeks after SB19's brush with infamy, executives at Constellation Brands were nonplussed when their popular lager, Corona Extra, became associated with a deadly viral strain that had begun to circulate in Wuhan, China. Corona is Latin for "crown," and that royal symbol serves as the beer's logo. It has been brewed and bottled in Mexico since the 1920s. Coronaviruses were identified in the 1960s and include a variety of pathogens, such as SARS and MERS. Their name is also derived from the Latin word for "crown": they have a spiky fringe that bears a passing resemblance to royal headgear.

It should be obvious that the beer and the virus have no relation to one another. That didn't stop people from wondering if there might be some connection, however. Searches for "beer coronavirus" and "corona beer virus" on Google surged by several thousand percent as the new and lethal virus entered public consciousness in mid-January 2020.[35] By late February, a poll found that 38 percent of Americans who drank beer would not buy Corona under any circumstances. Sixteen percent reported being uncertain about whether the lager was related to the virus.[36] To prevent such unintended associations, the new strain eventually received the moniker COVID-19. (Corona's sales rebounded sharply in the following months, as consuming alcohol became a popular way of coping with the pandemic and its lockdowns.)

It seems clear that boy bands and beer marketers should not be held responsible for inferences that the public might draw: after all, leaping to conclusions is a favored form of exercise in the online world. Other cases of unwarranted inferences may be more complex. Consider the phrase "chink in one's armor," which has been used in English since the fifteenth century to refer to a weakness or vulnerability. And "chink" itself (a fissure or a crack) has been attested by citations from as early as 1398. Shakespeare uses "chink in the wall" in *A Midsummer Night's Dream* in reference to two separated lovers.[37] Its use as an ethnic slur would come later, in the late nineteenth century. The offensiveness of the word is well known and has become so odious that most people avoid using it in any context.

In February 2012, an ESPN employee was fired for using the phrase "chink in the armor." Anthony Federico was an editor for the sports network, and he ran it as a headline in an article about Jeremy Lin. Lin was, at the time, a point guard for the New York Knicks and a major factor in helping his team reach the NBA playoffs that year—a period referred to by fans as "Linsanity."

Lin had been born in California in 1988 to parents who emigrated from Taiwan in the mid-1970s. Federico's use of the phrase, therefore, could be interpreted as a provocative racial epithet. He was, however, devastated when he realized what he had done, and the headline was removed from ESPN's mobile site half an hour after it first appeared. Federico was a fan of both Lin and the Knicks, and the headline he wrote was in reference to how Lin's nine turnovers contributed to the Knicks' loss to the New Orleans Hornets. He described it as an "honest mistake" and made an abject apology during an interview with the *Daily News*. Lin, for his part, was gracious in accepting Federico's apology.[38]

But the story doesn't begin or even end with Federico's dismissal. Earlier the same evening, ESPN anchor Max Bretos used the phrase in a live interview about Lin with Walt Frazier, the Knick's TV analyst. Bretos issued an apology on Twitter, pointing out that his wife was Asian and that he would "never intentionally say anything to disrespect her and that community." Bretos was more fortunate than Federico: instead of firing him, ESPN suspended him for 30 days.[39] And six years later, in 2018, TBS broadcaster Ron Darling referred to a chink in the armor of New York Yankees pitcher Masahiro Tanaka, who exhibited a loss of control early in a game. (Tanaka is Japanese.) Later, Darling issued a sincere and heartfelt apology.

Why do such incidents keep occurring? They may be the result of unconscious priming. In other words, the act of thinking about someone of Taiwanese or Japanese descent might cause the activation of a host of related concepts and schemas in long-term memory. These could include "Chinese," "Asian," "Orient," and, yes, "chink" and other racial slurs. Once such terms become activated, they are more accessible to mechanisms that govern language production and might be uttered or written by accident.[40]

At one time, such mistakes were referred to as Freudian slips: the Viennese psychoanalyst characterized them as overt manifestations of unconscious thoughts and wishes.[41] Rather than attributing such slips of the tongue to repressed desires, however, the mechanism of priming simply posits that certain terms will temporarily become more accessible. It's at least possible that Federico's headline was unconsciously primed by Lin's ancestry and by Bretos's earlier slip.

An episode similar to those involving Lin and Tanaka occurred in January 1995, when House Majority Leader Dick Armey (R-TX) referred to Representative Barney Frank (D-MA) as "Barney Fag."[42] Frank was an openly gay member of Congress, and Armey's awareness of this fact could have been the underlying reason that he uttered the derogatory epithet.

None of this should be construed as a defense for using such language. After all, most people know such disparaging terms and are nonetheless able to refrain from using them. An appeal to priming, however, can help us make sense of why individuals without a history of prejudice might say or write such things without thinking.

CAUTION: REPAIRS AHEAD

When miscommunication occurs during a conversation, who deserves the blame—the speaker or the hearer? And how do such problems get corrected? As it turns out, the answers to both of these questions depend on a variety of linguistic, cognitive, and social factors.[43]

In many cases, the miscommunication is triggered by the person who is doing the talking. She may misspeak, for example, by using the wrong word to refer to something. Or she may assume knowledge that is not possessed by her listener. Or her statement may be vague or ambiguous. In other cases, however, the hearer may be more culpable. He may not have been paying sufficient attention, or he may have misheard, misunderstood, or misinterpreted what the speaker was trying to say.

When the speaker is at fault, she has the luxury of correcting herself to head off any potential misunderstanding. After all, she still has the conversational floor and can make a quick attempt at amending her problematic utterance. For example, she may take another run at a word she has mispronounced or choose to elaborate if she sees confusion in the face of her listener. Language researchers refer to such episodes as repair sequences, and they have been studied extensively, both in the laboratory and in the wild, by sociolinguists, ethnographers, and cognitive scientists. As a result, we know a fair amount about how they work.

As we've already seen, repairs can be initiated by the speaker, but they can also be instigated by the hearer. And a repair can also be *completed* by either party.[44] A communicative rough patch, therefore, can be categorized as belonging to one of four groups. Let's consider each of these in turn.

In the case of self-initiated self-repairs, the speaker diagnoses her own conversational misadventure and fixes it without prompting from the hearer. Self-initiated self-repairs might be thought of as the conversational ideal: such sequences suggest careful monitoring by a speaker who is sensitive to any confusion on the part of her audience. An analysis of spontaneous self-repairs has found that they are typically self-interruptions, with an edit signaled by a particle like "uh" or "um," possibly followed by a pause as the speaker figures out the best way to clean up the mess she has made.[45]

In contrast, self-initiated other-repairs are diagnosed by the speaker but patched up by the hearer. For example, if the hearer knows the name of someone that the speaker is struggling to remember, he may interrupt to supply it, thereby facilitating the orderly progression of the conversation.

Other-initiated repairs occur when the listener breaks into the speaker's turn to signal confusion, disagreement, or misunderstanding. They are surprisingly common—a large, multilanguage study found that other-initiated repairs occur, on average, once every 84 seconds during conversation.[46]

In the case of other-initiated self-repairs, the hearer calls attention to a problem that is then corrected by the speaker. This may occur implicitly or explicitly. In face-to-face conversations, the speaker can monitor her partner for evidence of whether she is being understood. However, such signals can be ambiguous. A frown, for example, might be an expression of sympathy about the plight that the speaker is describing, but it can also function as an indicator of misunderstanding. A raised eyebrow might be a signal of incredulity, surprise, or confusion. And a pause on the part of the listener, before he begins his own contribution, may be the result of careful thought, surprise, or even disagreement with what has just been said.

Less ambiguous requests for repair may be initiated when the hearer asks a question. In English, the simple "huh?" or "what?" is usually enough to stop the speaker in her tracks, although some additional back-and-forth may be required to diagnose the source of the listener's confusion. Alternatively, the listener may be quite explicit by asking a question, such as "Do you mean the one on the right?"[47] Other-initiated requests are often marked by holds, in which "dynamic movements are temporarily and meaningfully held static."[48] Such holds typically involve eye gaze but can also involve the position of the eyebrows, head, or upper body.

There are, however, no guarantees that the speaker will diagnose and correct a communicative problem, even when confusion is signaled explicitly by the listener. Consider the following example from the TV sitcom *Seinfeld*, in which a stranger begins to converse with Elaine (Julia Louise Dreyfus) as they stand next to each other in a subway car:

Woman: I started riding these trains in the forties!

Elaine: Oh!

Woman: [to Elaine, but loudly] Those days a man would give up his seat for a woman. [softer, to Elaine] Now we're liberated; we have to stand.

Elaine: Hm. It's ironic.

Woman: What's ironic?

Elaine: This! That we've come all this way, we made all this progress, but . . . you know, we lost the little things, the niceties.

Woman: No, I-I mean, what does "ironic" mean?

Elaine: [confused] Oh . . .[49]

Elaine assumes that the concept of situational irony is universally understood and explains her point instead of the concept. As a result, this conversational interaction ends up going off the rails. Assumptions about shared common ground, as described earlier in this chapter, are a common cause of such derailments.

Finally, consider the case of other-initiated other-repairs. In these repair sequences, the listener does the heavy lifting by both diagnosing and attempting to fix the communicative problem. Repairs that are offered by the hearer can be problematic, however. For one thing, they are essentially an interruption, and the speaker may be less than thrilled about surrendering the conversational floor. In other cases, the hearer may be objecting to the speaker's version of events or her beliefs about something, which may lead to a disagreement.

Politeness norms within a given culture may dictate whether a listener will even attempt to repair a speaker's error. For example, he might realize that his partner has misspoken, but the error is a trivial one. In other cases, the intended meaning is transparent. In situations like these, it would be churlish to jump in with a correction.

And if there are differences in status between the conversational participants, a lower-status individual may be reluctant to correct someone of higher rank. A newly hired employee, for example, might be reluctant to correct the president of the company he works for—no matter how consequential the problem might be. All of this helps to explain why there is a bias for self-correction when conversational repairs are required and why other-initiated other-repairs are least common.[50]

3

PERCEPTUAL ISSUES

In this chapter, we examine the perceptual basis for miscommunication. Researchers have learned a great deal about how our minds make sense of the sounds we hear and the words we see. The ways in which our minds interpret this sensory input, however, can also lead to misperceptions and misunderstanding. We are so attuned to processing sounds as language that we are liable to do this even when we shouldn't. Similar issues arise with regard to the words that we read. And why is it next to impossible to proofread one's own writing?

DO YOU HEAR WHAT I HEAR?

In the late spring of 2018, the collective attention of the internet was briefly captured by an unlikely object of fascination: a one-second audio clip. It had been uploaded to Instagram by American student Katie Hetzel and went viral as it was reposted by others to Reddit and then Twitter. Online polls found that about half of those who listened to the recording thought they heard a male voice intoning the word "Laurel." Nearly as many, however, were convinced that they heard "Yanny" instead.[1]

For many, the controversy brought to mind a similar episode that occurred three years earlier. In that case, a photo of a dress was posted to Tumblr by Scottish singer Caitlin McNeill. Although the dress in the photo was blue and black, a substantial minority of viewers perceived it as white and gold instead. A scientific consensus has not yet emerged to explain such differing perceptions. It seems likely, however, that a variety of factors, such as chromatic adaptation and implicit assumptions about illumination, are responsible for the illusion of a white and gold dress.[2]

Is something similar going on with "Yanny or Laurel"? As with the dress, there is an underlying bedrock truth: the original recording had been made a decade earlier by opera singer Jay Aubrey Jones, who was contributing pronunciations to the website vocabulary.com. And the word that he produced and recorded was "Laurel." Why, then, were many people hearing what sounded like "Yanny"?

An important clue in cracking the case has to do with age: whereas perceptions of the dress are relatively uninfluenced by the age of the viewer, older hearers of the ambiguous audio clip tend to perceive it as "Laurel," whereas younger people are more likely to hear "Yanny." As it happens, sounds consistent with both interpretations are present in the recording: Jones's original had been rerecorded from the website by playing it through computer speakers and contains ambient sounds that added "noise" to the original. A perception of "Laurel" is consistent with attending to the lower-frequency sounds in the clip, whereas hearing the recording as "Yanny" is associated with the higher-pitched sounds.[3]

Age-related hearing loss, or presbycusis, is a nearly inevitable consequence of growing older. Although the decline in our ability to perceive sounds affects the entire audible range, the loss is greatest for higher-frequency sounds. Whereas people under the age of 20 can hear sounds up to about 19 kHz (kilohertz, or 19,000 cycles per second), those who are 30 may have trouble perceiving sounds with a frequency greater than 16 kHz. By the age of 40, this has typically dropped to about 15 kHz, and by age 50, the highest perceptible sounds may be only about 12 kHz.

This fact has been exploited by students to create "ultrasonic" ringtones, which their classmates can hear but their teachers cannot.[4] Shopkeepers have exploited this phenomenon in the opposite direction by installing so-called mosquito alarms that are marketed as acoustic deterrent devices. These gadgets emit high-frequency sounds (a typical value is 17.4 kHz) and are used to prevent younger people from loitering around storefronts or engaging in vandalism. Needless to say, their use has been controversial.[5]

Fortunately for those who are no longer young, the frequency range for human speech is far lower than for these mechanical noisemakers: between 125 Hz and 5 kHz. Speech sounds occupy a region in audiograms that is curved upward at both ends and is therefore sometimes referred to as the "speech banana."[6] Certain consonants, like "f," "s," and "th," occupy the upper end of this range. However, the harmonics, or overtones of speech sounds, extend to even higher reaches of the audible spectrum.

This can make it difficult for even middle-aged adults to distinguish between words that differ in such sounds, such as "fin" and "thin." The surrounding context can help to disambiguate between word candidates, but in noisy environments, in which several words or even entire phrases might be inaudible, the potential for misunderstanding and miscommunication increases rapidly.

It's important to realize that the consequences of hearing impairment affect communication in both directions. Although we typically think about its effect on impaired individuals, such as frustration, social isolation, and even cognitive decline,[7] their conversational partners are negatively affected as well. They may find themselves raising their voices—which in many cases doesn't help—or engaging in elderspeak, which is a simplified speech style that older adults often find patronizing.[8]

SIMILAR SOUNDS

The *Apollo 12* spacecraft was in serious trouble. Just 37 seconds after it had roared away from the Kennedy Space Center on November 14, 1969, the alarm lights on control panels in front of the three astronauts had lit up like a Christmas tree. On the ground, the flight controllers abruptly lost telemetry from the vehicle. After a textbook liftoff, it suddenly appeared as if the second flight to the moon might have to be aborted—a dangerous proposition even under the best of circumstances. (The cause of the malfunction was a power supply surge triggered by two lightning strikes, although this was not known at the time.)

The technicians in the Mission Operations Control Room in Houston tried to make sense of what had happened to the spacecraft, and time was of the essence. One of the flight controllers was a 26-year-old Oklahoman named John Aaron, who had been keeping track of the cabin pressure in the spacecraft as part of his electrical and environmental monitoring (EECOM) duties. Suddenly, his screen was full of nonsensical readings—a truly bad sign but also one that reminded him of an episode he had seen once before during a simulation.

The flight director, Gerry Griffin, asked Aaron for his input. He fully expected Aaron to recommend that the mission be aborted. Fortunately, Aaron was aware of an obscure setting that might restore the spacecraft's electrical system. If the signal conditioning equipment (SCE) were switched from its normal to its auxiliary setting, that might bring the capsule's electronics back to life.

In response to Griffin's query and 14 seconds after the first lightning strike, Aaron said, "Flight, EECOM. Try SCE to aux."

Griffin, who was puzzled by this esoteric command, asked, "Say again, SCE to off?" He had misheard Aaron's response:

Aaron: Aux.

Griffin: SCE to aux?

Aaron: Auxiliary, Flight.

Griffin: SCE to aux, Capcom.

The Capcom, or capsule communicator—the person who talked to the astronauts—was Jerry Carr. He was also puzzled by Griffin's reference to the obscure switch but relayed the command:

Carr: *Apollo 12*, Houston. Try SCE to auxiliary.

On *Apollo 12*, it was Commander Pete Conrad's turn to be baffled:

Conrad: NCE to auxiliary? What the hell is that? [He had misheard Carr's request.]

Aaron (urgently, to Carr): SCE. SCE to auxiliary.

Carr (to Conrad, enunciating each letter distinctly): S. C. E.

Conrad (to his crew): SCE to aux.

Fortunately for everyone involved, Lunar Module pilot Alan Bean knew the location of the SCE switch among the forest of controls in front of him, and he toggled it to the auxiliary position. Suddenly, telemetry data began to flow once again to Mission Control. The moon mission had been saved.[9]

In many ways, ground communication with an *Apollo* spacecraft was a real-life version of the Telephone game: the flight controller had to relay his instructions to the flight director, who relayed them to the capsule communicator, who spoke to the mission commander, who passed them to his crewmates. Such chains are only as strong as their weakest link, and adding more links only compounds the problem. In addition, the flight controllers and directors used a bevy of acronyms, initialisms, and abbreviations to make communication as efficient as possible. And all of this unfolded in a noisy environment via low-fidelity communication links that were prone to static and dropouts.

In the exchange between the spacecraft controllers and crew, we see several instances of miscommunication and misunderstanding. Not only was Aaron proposing what seemed like an unusual solution to the craft's electrical problem, but the language itself was getting in the way. The problem was that some of the critical words and abbreviations employed by the controllers were very much alike.

We can see this in Griffin's initial response to Aaron's recommendation: "Say again, SCE to off?" Griffin mishears "aux" as "off," which is clearly a possible setting for a switch but not the right one. Aaron corrects him by repeating the intended word, and Griffin repeats it as well to check his understanding. At this point, Aaron uses the full term "auxiliary" to allay any remaining confusion about his proposed instruction for the crew.

A similar problem occurs when the Capcom tries to relay the instruction to the commander of the spacecraft. Carr tells Conrad, "Try SCE to auxiliary," making a point of using the full term to minimize any confusion about the setting for the switch. However, Conrad mishears "SCE" as "NCE," which is made apparent when he repeats the command and expresses his lack of understanding. Carr has to repeat each letter of the control separately and distinctly so that Conrad finally understands what is being asked of him and his crew.

At a perceptual level, such miscommunication can occur because consonant sounds have most of their acoustic energy at relatively higher frequencies in comparison to vowel sounds. This makes them harder to distinguish when heard through tinny speakers or in noisy environments.[10] As a result, the final sound in "aux" is hard to distinguish from the final sound in "off." (If you've ever resorted to spelling out an abbreviation over a bad cell phone connection—"'F' as in 'Frank,'" "'S' as in 'Sam,'" and so on—then you've also experienced this problem.) In short, the use of low-fidelity devices, whether on Earth or in space, counts as one strike against successful communication at the outset. And if ambiguity is added in, the likelihood of a communication failure is high.

The flight controllers, directors, and crew of *Apollo 12* had successfully diagnosed and fixed a serious and unanticipated problem. Five days later, Pete Conrad and Alan Bean landed at the Ocean of Storms and spent nearly eight hours exploring the lunar surface. A mission that had almost ended as soon as it began was rescued by the quick thinking of John Aaron and his Mission Control colleagues. Throughout this flight and many others, they provided direction using arcane terminology and commands that could be misheard, misinterpreted, and misunderstood. To their credit, they were well aware of the possibility of such problems and followed protocols to minimize miscommunication. The tedious repetition of commands before they were enacted, for example, might seem unnecessary, but as the "Try SCE to aux" exchange demonstrates, they were absolutely essential for the success of the mission.

HEARING VOICES

Conspiracy theorists have found a congenial home on contemporary social media sites. But decades before QAnon, birthers, and truthers, a baroque fantasy was being spread about the Beatles. Specifically, it was claimed that Paul McCartney had died in an automobile accident in 1966 and that he had been replaced in the British band by a look-alike. Evidence for "Paul is dead" could be found, it was said, in subtle clues on the band's album covers and in the lyrics of their songs.

The example cited most often can be found in "Revolution 9" from the band's 1968 *White Album*. When the phrase "number nine" is played backwards, it sounds—a bit—like "turn me on, dead man."[11] The dead man in question would be Paul.

The technique of intentionally recording a message backward is called backmasking. The Beatles did use backmasking in the fade-out of "Rain," a song on 1965's *Rubber Soul* album.[12] Therefore, according to this tortured logic, it was possible that the band was using backmasking to send out subtle clues to the faithful about McCartney's demise. (Those who wish to hear the forward and backward versions of the putative message can find them in the "Paul is Dead" article on Wikipedia.)

There are several problems with all of this. One is that McCartney, at the time of this writing, is still alive at age 80, and recently completed a tour of the United States. He has released two dozen albums with his group Wings as well as solo work since his supposed death more than 50 years ago. The other problem is that the backward version of "number nine" sounds like "turn me on, dead man" only if that is the phrase one expects to hear.

The human mind has evolved an exquisite ability to pick out patterns and to separate signal from noise. Sometimes, however, it does its job too well and finds meaningful patterns where none exist. This is referred to as pareidolia,[13] and examples include finding animals in clouds, faces on the fronts of cars, a man in the moon, or even canals on Mars—an example we will return to in chapter 9. And pareidolia also applies to ambiguous strings of sound, like backward language.

But what if there is no external signal at all? Some people hear voices when no one else is present. Many associate this phenomenon with mental illnesses, such as schizophrenia. And although hearing voices is a characteristic symptom of that disorder, the incidence of schizophrenia is relatively low: the overall rate is probably between 1 and 2 percent.[14] On the other hand, about 10 percent of people report having heard voices, although estimates vary widely. The rate is slightly higher for children and adolescents and lower for younger and older adults.[15] This suggests that many people who are psychologically healthy experience what researchers refer to as auditory verbal hallucinations (AVH).

A distinction should be made between people who experience AVH rarely and those who experience the phenomenon frequently. Hearing someone speak your name when no one is present may have a distinctly different cause than those who experience such hallucinations on a regular basis and engage in conversations or arguments with the phantom voices.

The fact that people tend to hear voices as opposed to, for example, running water or other environmental sounds may reflect our predisposition to interpret ambiguous sounds as speech. Researchers have found, for example, that people who report hearing voices are better able to detect ambiguous or degraded speech than control participants who do not experience AVH.[16] In other words, the brains of people experiencing AVH perceive *something* and attempt to impose some sort of order on it. The stimulus may be external, but it can also be triggered by spontaneous or random neural activity in the left temporal lobe of the brain.[17]

Psychologists have long known that the perceptual threshold for hearing one's own name is lower than it is for other, more neutral stimuli.[18] This may explain why we tend to hear extremely familiar words, such as our own name, instead of, say, "asparagus" or "calliope" when the brain tries to make sense of ambiguous ambient sounds. In his book on hallucinations, the neurologist Oliver Sacks suggests another possibility: it may be that the brain fails to differentiate between internally generated speech—such as when we talk to ourselves—and externally generated language.[19]

The list of famous individuals who report having heard voices, at least during some part of their lives, is long and varied. Musicians are well represented: Brian Wilson of the Beach Boys and Lady Gaga would be examples, as are actors, such as Anthony Hopkins; authors, including Philip K. Dick and Charles Dickens; and philosophers, such as Swedenborg and Socrates. Psychiatrists, who may be particularly attuned to their internal states, such as Sigmund Freud and Carl Jung, also reported experiencing AVHs. And many religious figures, such as Moses and Joan of Arc, have interpreted disembodied voices as directives from God.

Explanations for AVH run the gamut from genetic to environmental factors, with the latter including anxiety and prior trauma. A combination of stress and high caffeine intake, for example, caused some research participants to hear the song "White Christmas" while listening to random noise. In that study, the song was never actually played, but individuals who were in a low-stress condition—and who drank less coffee—were significantly less likely to experience this particular halluci- nation.[20] Perhaps it was stressed-out and overcaffeinated college students in the 1960s who convinced themselves that Paul McCartney had joined the choir invisible.

LADY MONDEGREEN AND SLIPS OF THE EAR

Almost everyone has discovered, at one time or another, that they were mistaken about the lyrics of a particular song. In an episode of the sitcom *The King of Queens*, actor Kevin James begins to croon Neil Diamond's song about a stylish man of the cloth: Reverend Blue Jeans. His long- suffering wife, played by Leah Remini, has to point out that the phrase is "forever in blue jeans," much to her husband's consternation.

However, it's not only TV characters who make such mistakes. And a term exists for such mishearings: they're called Mondegreens. This unlikely moniker was coined by the American author Sylvia Wright. In an essay she wrote in 1954, Wright recalled her mother reading to her when she was a child. A favorite was the Scottish ballad "The Earl of Moray." The first verse contains the lines "They have slain the Earl of Moray/And laid him on the green." Wright misheard the second line as "And the Lady Mondegreen" and imagined the slain couple arranged side by side. Not finding a term to describe this phenomenon, Wright created her own, and such slips of the ear have been referred to as Mon- degreens ever since.[21]

Examples of such errors are legion. Well-known specimens can be found originating in the lyrics of Jimi Hendrix ("Excuse me while I kiss this guy" instead of "the sky"), the Beatles ("the girl with colitis goes by" instead of "kaleidoscope eyes"), and Prince ("rats marry their friends" for "raspberry beret"). The author of this book thought that, in their song "Africa," Toto was speculating that "I guess it rains down in Africa." The correct version is "I bless the rains." He has heard this song many times since its release in 1982, when he was in college, and discovered his mistake only recently, nearly 40 years later. It almost goes without saying that a website has sprung up to collect and archive such Mondegreens, and it can be found at kissthisguy.com.

Such mishearings may be amusing, but they also tell us a great deal about how people perceive—and misperceive—speech. Almost by definition, the acoustic signal that reaches our ears is ambiguous, and in many cases, the same stream of sound can be interpreted by the brain in more than one way. This is particularly true for songs, in which the speech signal is particularly noisy: there are other sounds—the musical accompaniment—competing with the words, and the lyrics are being sung as opposed to spoken, which can subtly affect the prosody of the words. And unless the listener is hearing the music live or watching a video, she doesn't have access to cues from the singer's mouth movements, which can provide disambiguating information.[22]

Mondegreens probably have a different source than malapropisms, which are word substitution errors. These are named after a character in *The Rivals* (1775), a work by the Irish playwright Richard Sheridan. In that play, a Mrs. Malaprop is the guardian for one of the main characters, Lydia Languish. She makes several of these mistakes, such as encouraging Lydia to forget and "illiterate" a particular suitor from her memory or despairing that Lydia is as headstrong as an "allegory on the banks of the Nile." (Shakespeare also made use of such errors for comic effect, as in some of Dogberry's dialogue in *Much Ado About Nothing*. Many of the contorted observations of Yogi Berra, such as acknowledging that "Texas has a lot of electrical votes," would qualify as well.)

Researchers believe that the underlying problem in the case of malapropisms has to do with word retrieval: the speaker may accidently initiate the motor program for articulating a different word with a similar pronunciation. At least in some cases, Mondegreens and malapropisms may have a similar origin in that an unfamiliar word or phrase is misheard and remembered incorrectly.

In other cases, it may be a person's accent or dialect that gets in the way. The American linguist Meghan Sumner remembers ordering a sliced turkey sandwich at a deli on Long Island and being startled when she was offered a choice between white meat and "dog meat."[23]

It is interesting to note that, for both Mondegreens and malapropisms, the misheard or misspoken word or phrase is often more familiar than the original. "Kaleidoscope eyes" or "raspberry berets" are not expressions that are especially common, and neither are "obliterate"

(Mrs. Malaprop) or "electoral" (Mr. Berra). Sylvia Wright, as a child, may have never heard of a village green. And such mishearings can also occur crosslinguistically. It is not unusual for second-language learners to mistakenly hear words from their native language in the songs of their newly acquired tongue.[24] It appears, therefore, that our minds prefer hearing the customary and the routine, and we are liable to misinterpret the ambiguous or the unfamiliar as a result.

This process of normalization also seems to operate at the level of society, as rare and unusual words are replaced, throughout time, by soundalike equivalents. The expression "champing at the bit" has, for many, been transformed into "chomping at the bit." Although the word "champ," in the sense of chewing, has been part of English since at least 1530, it has clearly fallen out of favor in comparison to "chomp." Shakespeare's aphorism from *The Merchant of Venice*, "All that glisters is not gold," is now more commonly expressed with the word "glitters." And the title of Thomas Hardy's 1874 novel *Far from the Madding Crowd* is sometimes mistakenly referred to as "far from the maddening crowd."

A year before his death, the neurologist and author Oliver Sacks published an article about his own slips of the ear in the *New York Times*. He related his surprise when his assistant announced she was leaving for choir practice, even though she had never mentioned such an interest previously. As it turned out, she was off to see her chiropractor. On a later occasion, when she humorously announced she was departing for choir practice, he heard "firecrackers" instead. Sacks attributed such mishearings to his increasing deafness and decided to keep a journal of such episodes. Although they were clearly errors, Sacks viewed them as products of his linguistic ingenuity and seemed fascinated by his mind's ability to manufacture such alternative interpretations.[25]

Once again, we see how a perceptual deficit, like being hard of hearing, counts as a strike against successful communication. And when this impairment is paired with a low-frequency term, like "chiropractor," the system breaks—in this case by proposing imaginative albeit incorrect interpretations of the imperfectly heard string of sounds.

SEEING IS (NOT) BELIEVING

In a now classic psychology experiment, J. Don Read asked people how many instances of the letter F appear in the sentence below. Give it a try before reading further:

FINISHED FILES ARE THE
RESULT OF YEARS OF SCIENTIFIC
STUDY COMBINED WITH THE
EXPERIENCE OF YEARS.

So, how many did you find? If you're like the vast majority of people, you probably spotted only three Fs. When Read conducted his study, he found that between 85 and 90 percent of the participants found only three instances.[26] In reality, however, the sentence contains six. The Fs in "finished," "files," and "scientific" are relatively easy to spot, whereas the Fs in the three instances of "of" are much harder to detect. Why might this be?

Read proposed that letter detection failures can be explained by the way in which the mind processes text. The idea is that during the act of reading, words are converted into strings of sounds—a process called phonological recoding.[27] The words "finished," "files," and "scientific" contain sounds that map onto the "f" sound, whereas the Fs in "of" sound like Vs. And if we are on the lookout for Fs, we may fail to notice them in words that don't contain the "f" sound but do contain the letter F.

When we think about what the act of reading feels like, it seems a lot like mowing the lawn. Much as we methodically march back and forth across a yard, our eyes seem to smoothly swivel along a row of letters from beginning to end before jumping to the next line to repeat the process. This intuition, however, doesn't square with reality. Our eyes hop along lines of text, and not every word is directly gazed on. These hops—technically referred to as saccades—cause us to skip short, common words that can be readily inferred from context.[28] This phenomenon helps to explain why we fail to notice the F in frequently used words like "of" but do spot them in longer, less common words like "scientific."

In addition, the act of reading might be better described as the act of prediction. As our eyes hop along a line of text, the mind is busy generating inferences and hypotheses about what happens next.[29] It's easy to catch this process in action when our expectations are violated.

If, for example, we read "Two men walk into a bar," we may recognize this as the beginning of a joke and may even start generating inferences about what might happen next—an interaction with the bartender perhaps. However, if the next sentence is "The third one ducks," most of us will do a double take. The first sentence turns out not to be the setup for a joke, and the bar has to be reconceptualized as a long rod instead of an establishment serving alcohol.[30] As we will see with garden path sentences in chapter 7, our predilection for prediction sometimes leads us astray.

If we put the eye hopping and the predicting together, it becomes easier to understand why most of us are bad at proofreading. No matter how many times we scan through what we've written, we are likely to miss spotting many of the misspelled words as well as missing or doubled words. Research has shown, for example, that when asked to identify specific words in a text, people perform remarkably poorly when asked to find instances of short function words like "the" or "an"—their eyes hop right over them.[31]

At this point, you might be thinking that technology has solved such problems. After all, doesn't software like Microsoft Word flag spelling and grammar errors? It is true that computer programs can identify and even correct many mistakes that humans might not see. For example, people spot repetitions of the word "the" (as in "She jumped off the the board and into the water") only about half the time.[32] Simple computer algorithms, on the other hand, will identify all such instances.

For more complex cases, however, relying on software to identify errors is problematic at best. For example, if you were to type the phrase "Noah built a a huge ark," Microsoft Word would obligingly spot an error and place a red squiggle under the second, superfluous article. But the program won't bat an eye if you type "Noah built a huge arch" or, for that matter, "Moses built a huge ark." Spelling and grammar checkers aren't sophisticated enough—yet—to detect such mistakes because they require inferences or world knowledge that are difficult to translate into error-checking code.

And if reading is predicting, this suggests that we should be especially bad at proofreading text that we have written ourselves. After all, we are already familiar with the content of what we've composed, and this causes the prediction process—generating what comes next—to be turbocharged. Experimental work supports this idea: research participants found it easier to detect errors in the compositions of others when compared to finding problems in their own writing. However, after a two-week interval, this difference largely vanishes.[33]

This supports the conventional wisdom that it's easier to find the errors in one's own writing if a composition is put aside and returned to later. The passing of time will lessen the familiarity of what was written, and the errors will be easier to spot and correct.

One of the best ways to proofread a document is to have a computer read it aloud to you. Reading it aloud to oneself isn't terribly effective because your eyes will still fail to spot errors, and your mind will fill in the missing words, causing your mouth to speak them. In contrast, a program like Microsoft Word will read aloud exactly what you have written—including all of the infelicities—and this makes it easy to hear and correct them.

THE CURSE OF CURSIVE

As we have already seen, miscommunication has a variety of causes and consequences. Some of these outcomes, however, are more consequential than others. In the field of medicine, for example, the effects can be deadly.

In July 2006, the National Academy of Science's Institute of Medicine estimated that preventable errors were injuring more than 1.5 million Americans every year. The reasons for these mistakes included incorrect dosages, cryptic abbreviations—and sloppy handwriting. The institute estimated that illegible scrawls by physicians resulted in the deaths of more than 7,000 people each year.[34]

As an example, consider the case of a 42-year-old patient from Texas who was prescribed 20 milligrams of Isordil, to be taken every six hours, for angina. Because of his cardiologist's poor penmanship, however, the pharmacist misread the drug as Plendil, which has a maximum dosage of 10 milligrams daily. The overdose led to a fatal heart attack, and a jury found both the prescribing physician and the pharmacist liable for the deadly mistake.[35]

If the names of the thousands of trade and generic drugs were distinctly different from one another, the correct name could probably be deduced from even a hastily written scrip. However, many medications have similar names, as with the two drugs ending in "dil" in the last paragraph. These "soundalike" drug names are responsible for a up to a quarter of all medication errors.[36]

Fortunately, this problem is being addressed through technology. Most hospitals have instituted digital systems such as the Computerized Physician Order Entry (CPOE). In such systems, prescriptions are entered via a keyboard and transmitted electronically to a pharmacy. The adoption of electronic prescribing has led to a significant reduction in medication errors since these systems came into wide use.[37]

Although problems with physicians' poor penmanship have been addressed, another handwriting dilemma has taken its place. Many school systems have stopped teaching cursive handwriting altogether. In these jurisdictions, cursive has been squeezed out of the curriculum by competing demands on teachers' time, such as preparing for mandatory state testing. And its omission from the Common Core curriculum in 2010, in favor of teaching keyboarding skills, seemed to sound the death knell for supporters of cursive instruction.[38]

It could be argued that cursive handwriting has outlived its usefulness and should take its place in the pantheon of skills rendered obsolete by technology.[39] To cite but one example, it has been decades since most high schools or vocational institutions taught Pitman's or Gregg's shorthand. The need for rapid note taking was largely replaced by a steady march of alternatives, such as typewriters, dictation machines, and word processors. And should someone need to write a thank-you note, they can always fall back on printing in block letters. It may not be elegant, but it is serviceable enough.

By the mid-2010s, however, there was legislative pushback in more than a dozen states to mandate the teaching of cursive, and it has now been reintroduced into the education standards of many states.[40]

Although the pendulum seems to be swinging back in cursive's favor, some millennials are entering the job market without the ability to read or to write cursive handwriting. For them, the issue is not being able to decipher illegible handwriting but rather the inability to read script of any type.

Illegibly written addresses have long been the bane of the post office, which at one time employed thousands of workers to decipher sloppily written or otherwise indecipherable numbers, streets, and city names. Once again, technology has largely solved this problem as optical scanning machines improved in accuracy. The U.S. Postal Service once maintained 55 centers for such work, but by 2013, only one remained.[41]

Does sloppy handwriting indicate anything more than poor penmanship? Might it, for example, reveal characteristics of the writer's personality? A belief in graphology, as it is formally known, is taken seriously by some corporations that employ handwriting analysts to provide input on personnel recruitment and selection.[42] One guide to the field characterizes illegible handwriting as a characteristic seen among artists and also in "unconventional characters whose inner life is so strongly developed that they completely lose sight of the world around them."[43] Although the field has its defenders in some quarters, experimental studies have consistently failed to find an association between handwriting, measures of personality, and on-the-job behavior.[44]

As it turns out, however, handwriting does have some predictive value: it can be a harbinger of disease. Specifically, changes in one's handwriting have been reliably linked with the development of Parkinson's disease. This degenerative disorder of the central nervous system causes profound changes in an individual's motor control, with depression and dementia becoming common as the disease progresses. A symptom that is frequently seen in the early development of the disorder is micrographia: the handwriting produced by the afflicted individual becomes smaller and more cramped.[45]

4

WORDS, PART 1

Although psychological and perceptual factors are important, miscommunication isn't all in your head. In some cases, problems are caused by the sound structure of the language, or by differing ideas about what particular words refer to.

HARD TO SAY

> *Then said they unto him, Say now Shibboleth: and he said Sibboleth: for he could not frame to pronounce it right. Then they took him, and slew him at the passages of Jordan: and there fell at that time of the Ephraimites forty and two thousand.*
>
> —Judges 12:6, King James version

On August 4, 2020, Donald Trump gave a speech on the occasion of his signing the Great American Outdoors Act. "We want every American child to have access to pristine outdoor spaces," he intoned, reading carefully from the typescript he was holding. "Where young Americans experience the breathtaking beauty of the Grand Canyon, when their eyes widen in amazement as Old Faithful bursts into the sky." Warming to his task, the president injected a note of wonderment into his last few words.

But he stumbled when tackling the next clause in the speech: "When they gaze upon Yosemite's . . . Yosemite's, towering sequoias." Instead of referring to the national park in California as "yo-SEM-it-ee," however, he said something like "Yo-SEM-ight" and then tried again with "Yo-SEM-in-ite." People took to Twitter to mock the president's mangling of the word, suggesting that perhaps he meant to beckon a person of the Jewish faith instead.

Had Trump never heard this word before? It seems unlikely. He is undoubtedly familiar, for example, with the Looney Tunes cartoon character Yosemite Sam, the short-tempered and trigger-happy cowboy whose temperament closely resembles Trump's own. And the president's difficulties with the word were reminiscent of Kelly Bundy's, the ditzy teenager from the TV sitcom *Married . . . with Children*, who in one episode pronounced the word as "YO-ze-mite."

Trump's wrestling match with Yosemite wasn't a one-time problem. He has struggled with other words, such as "chasms" (employing a "ch" sound) and "Nazis" (using a "z" instead of the German "ts" sound), or stressed the wrong syllable in the names of countries, like referring to the East African nation Tanzania as "Tan-ZAY-nia."

It's a problem common among people who don't read much—a fact the president has asserted on a number of occasions, claiming he's too busy to read.[1] In the case of Yosemite, he was simply unable to map the pronunciation of a word he knew to its appearance on the page. Instead, he tried to sound it out as if it were a phonetically regular English word, with a silent "e" marking a lengthened /i/, as in "dynamite," "stalagmite," and, yes, "Semite." The problem is that the name "Yosemite" isn't English. It's from the Miwok language, which was spoken by a Native American tribe that once lived in the Yosemite Valley.

Trump isn't the first American president to grapple with words that can be difficult to pronounce. On June 26, 1963, John F. Kennedy gave a speech in West Berlin, in front of the notorious Wall less than two years old, at a time when Cold War tensions were high. Wanting to emphasize solidarity with his audience, JFK included a bit of German in a last-minute addition to his speech. "Two thousand years ago, the proudest boast was 'Civis Romanus sum' [I am a Roman citizen]. Today, in the world of freedom, the proudest boast is 'Ich bin ein Berliner.'"

Kennedy, who was "notoriously tongue-tied when it came to foreign languages,"[2] took care to avoid mangling the phrase. His handwritten notes from the address have been preserved and show that he carefully spelled out the words phonetically: "Ish bin ein Bearleener." (The claim that it means "I am a jelly doughnut" is an urban legend that would gain currency many years later.[3]) Trump would have been well advised to follow Kennedy's example in his own remarks about national parks.

In some cases of mispronunciation, a preferred version may not be widely known. Many political candidates and news anchors stumble over the pronunciation of "Nevada." It's often pronounced like "ne-VAH-duh," although residents of the Silver State will tell you that they pronounce it "Ne-VAD-uh." Their preference didn't keep candidate Trump from declaring, at a 2016 rally in Reno, that the state's inhabitants were pronouncing it wrong.[4]

The residents of Florida, Missouri, and other states also have preferred pronunciations that don't always square with the intuitions of those hailing from elsewhere. Getting a name wrong can have consequences since such mistakes mark someone as an outsider who may not be completely trustworthy. That's not a good look for anyone, especially for a politician.

Another example that bedevils many Americans is the place name "Worcestershire." Ordinarily, this would be an issue only for tourists visiting the English West Midlands, who would discover that the locals compress it into three syllables, rendering it something like "WUS-te-sher." The place name, however, is also found in Worcestershire sauce, the much-loved fermented condiment. As a result, those who like to spice up their Bloody Marys or deviled eggs with it must grapple with the large variance between its spelling and pronunciation.

In other cases, mispronunciations result from a lack of familiarity and a bad case of nerves. An embarrassing example occurred in 2013 on the syndicated game show *Wheel of Fortune*. One of the contestants, Paul Atkinson, was understandably nervous. The firefighter from Oregon was taking his first step toward a chance at winning a million dollars. All he had to do was complete the phrase "Corner curio ca——inet." Unfortunately, he only managed to blurt out something like "cornoh curl cabinet." Pat Sajak, the show's host, ruled against him for mangling the phrase, and Atkinson ended up leaving with only $2,000.[5]

When interviewed later, he admitted to never having seen the word "curio" before, and this, combined with his nervousness, led to the mispronunciation. It wasn't a total loss, however. Atkinson's notoriety led to an appearance on *Late Night with Jimmy Fallon*, where he was presented with—you guessed it—a lovely curio cabinet.

Sometimes, the language itself is to blame. English has many words with consonant clusters that can be hard to articulate cleanly, such as "desks" and "mists." And the so-called rhotic /r/ can be challenging when repeated: a running joke on the TV comedy *30 Rock* was the difficulty in pronouncing the phrase "the rural juror," which was the name of one of the character's films.

Foreign names are even more problematic since they may make use of sounds or sound combinations that aren't employed in a speaker's native language. During both world wars, many American soldiers were flummoxed by the pronunciation of the names of French cities and towns. To cope, they simply made up their own versions: "Béziers" was transformed into "brassieres," and "Sainte-Mère-Église" became "Saint Mare Eggles." Andy Rooney, who served as a war correspondent, opined that Reims was a poor choice for the location of the German surrender in 1945 because it was virtually impossible for anyone but the French to pronounce correctly.[6]

OF HOMOPHONES AND EGGCORNS

> *Why was the nighttime sad?*
> *Because it was mourning.*

> —Children's riddle

> *Reagan Goes for Juggler in Midwest.*

> —Headline, *Charleston Gazette*, 1984

In the crackdown that followed the Tiananmen Square protests in 1989, Chinese police were on the lookout for subversive activity and in particular any criticism of the Communist government. In one particularly egregious example, a group of police severely beat a man for dropping a bottle, which they interpreted as criticism of the country's leader, Deng Xiaoping.[7] Although it may be hard to see the connection between breaking a bottle and a political act, a brief digression into the sounds and meanings of words may prove helpful in making this connection.

In some ideal language, every string of sounds and letters would map on to one referent and one referent only. This is not the case in most if not all of the world's languages, and it creates a phenomenon that researchers have dubbed lexical ambiguity. There are several different varieties, depending on whether the spelling, pronunciation, or the meaning of two words are the same or different.

Let's tackle sound differences first. Homographs are words that are spelled the same but are pronounced differently. An example would be "lead." When pronounced with a short vowel, it refers to the chemical element. However, when the same word is spoken with a long vowel, it refers to the act of guiding someone. In other cases, a change in stress can change the meaning, as in "desert." When the first syllable receives stress, the word refers to a sandy wasteland. When the second syllable gets the emphasis instead, it denotes abandoning something.

In other cases, a word might be spelled the same and pronounced differently and yet still mean the same thing. This is true across the various dialects and accents of a given language. For example, Americans and Canadians pronounce "about" differently, but the word still refers to the concept of approximation (as in "about an hour," which is a near rhyme in Topeka but not in Toronto).

Homonyms are words that have the same spelling and pronunciation but mean different things, such as the river and money senses of "bank" or the animal and carry senses of "bear." A primary task for lexicographers is carefully distinguishing among the various senses of a given word. Some terms have dozens: the *Oxford English Dictionary* lists nearly 100 senses for the word "set" (47 meanings as a noun, 44 as a verb, and seven as an adjective).

Finally, homophones are words that have different spellings and meanings but are pronounced the same, as in the English homophonic triplet of "to," "too," and "two." They are often used in puns or other forms of simple wordplay, as in the children's riddle at the beginning of this section.

It's even possible for two words that are pronounced identically to have contradictory meanings, as in "raise" (to lift up) and "raze" (to tear down). There are certainly contexts in which this could create ambiguity, as in hearing the sentence "The construction crew raised/razed the old building." In such cases, one word in the pair seems to be dominant with regard to usage: "raze" is a relatively uncommon English word, whereas "raise" is common. If both were used frequently, confusion could result.

Homophony is significant in many non-Western languages as well. A good example is Mandarin, which is based on the dialect of Chinese spoken in Beijing. Most words consist of only one or two syllables, and they can be uttered using one of four tones.

Mandarin has many homophones: the words for "star" and "gorilla," for example, are pronounced the same, both in sound and in tone, although they are written with different characters. The language also has many homographs: the same monosyllable, such as *shi*, can mean "lion," "ten," "history," or "to be," depending on the tone employed by the speaker.

This brings us full circle to the dropping of a bottle as a form of protest. The Chinese leader's name, "Xiaoping," and "little bottle" are homophonic in Mandarin. And to express their disapproval of Deng, thousands of dissidents smashed glass bottles in public as an indirect criticism of the Chinese government and its premier leader.

For those whose primary exposure to language is through its spoken rather than written form, it's not hard to imagine that many familiar phrases are misheard and learned as homophonic equivalents.[8] And when such individuals express themselves in writing, they may reveal their creative mishearings through such inventive expressions as "for all intensive purposes," "pass mustard," "taken for granite," and "pinochle of success." Such substitutions were dubbed "eggcorns" (a mishearing of "acorn") by linguist Geoffrey Pullum in 2003.

The linguist Mark Liberman has argued that eggcorns deserve their own name because they are distinctly different from malapropisms, a form of mishearing we encountered in the last chapter. In the case of

eggcorns, the mishearing is homophonic or almost identical in sound—
unlike Mrs. Malaprop's substitution of "allegory" for "alligator."

Eggcorns also aren't Mondegreens because they aren't part of a poem
or a song. And they also don't qualify as folk etymologies, which are usually shared by a large number of people.[9] It should come as no surprise to
learn that a collection of several hundred such expressions can be found
online at eggcorns.lascribe.net.

The editors at Merriam-Webster enshrined the neologism "eggcorn"
in their dictionary in 2015.[10] Eggcorns don't appear in print often
because vigilant editors spot and correct them. But some do slip past,
as in the headline at the beginning of this section. The phrase creates a
vivid mental image of the former president pursuing a hapless carney, but
it's probably not what the journalist intended to convey.[11] Eggcorns have
probably existed for centuries if not millennia and revealed themselves
only as literacy became widespread.

ON THE CONTRARY

As discussed in the previous section, many words are homonyms; that is,
they have more than one meaning. And in most cases, it's unlikely that
these multiple senses will create confusion. For example, "bear" can refer
to the ursine animal or to the act of carrying something. These senses are
distinctly different, and the context in which they're employed will usually be sufficient for arriving at the intended meaning.

Some words, however, have senses that are in opposition to one
another. Consider the word "sanction." It can mean to approve something, as in "The United Nations sanctioned the group." However, it can
also mean imposing a penalty, and the example in the previous sentence
works for that case as well. Context may not always help in such cases
since an organization like the United Nations may have the power to both
approve and to penalize. Such words, therefore, create yet another type
of ambiguity and potential for confusion.

Is it common for words to have such diametrically opposed meanings?
Contronyms, as they're called, pop up frequently enough to be problematic.[12] They were named by Jack Herring in 1962[13] and are also referred

to as auto-antonyms and as Janus words: like the two-faced Roman god, these words are their own opposites.

Examples of contronyms include "clip," which can mean to fasten together or to detach. "Transparent" can mean invisible or obvious. "Refrain" can mean to not do something or to repeat it. To "dust" can mean to add fine particles or to remove them. English has dozens of such meaning pairs, and they occur in other languages as well.

It should be noted that although there are many contronyms, some of them involve relatively obscure words. The verb "cleave" is often cited as an example, but neither of its two, diametrically opposed meanings (to split apart or to stick together) could be described as being common in contemporary usage. The same is true for "liege": it can refer to a feudal lord or to a vassal, but such distinctions are unlikely to pop up in casual conversation.

Perhaps the most controversial contronym is "literally," which can mean both something that is objective and factual as well as something that is nonliteral or figurative (which is the literal opposite of its literal sense). In its figurative sense, "literally" is used as a hyperbolic intensifier, as in "I literally moved heaven and earth to get to my meeting on time."

Many language purists have been outraged to discover that this second, patently contradictory sense has been enshrined in dictionaries and are under the impression that this addition occurred fairly recently. The lexicographers at Merriam-Webster have responded to this criticism by pointing out that "literally" in its figurative sense is not some contemporary debasement. On the contrary, it was employed in this way as early as 1769. Authors who have used "literally" in its figurative sense include Charlotte Brontë, Charles Dickens, James Joyce, and William Makepeace Thackeray. In addition, this sense has been listed in Merriam-Webster dictionaries since 1909.[14]

In some cases, a contronym's meanings are closely related, and it becomes virtually impossible to figure out what is meant. The poster child in this case is probably the word "peruse," which can mean either to read carefully or to skim casually. Benjamin Dreyer, in his delightfully opinionated style guide, proclaims that "peruse" "is as close to useless as a word can be."[15]

As we saw in the previous section, some homophones can be contronymic as well, such as "raise" (to lift up) and "raze" (to tear down). To avoid confusion, a kind of Darwinian struggle may, throughout time, relegate one member of the pair to relative linguistic obscurity.

Even near contronyms can be problematic and can create yet another obstacle in the path of second-language learning. For example, the German words for "buy" and "sell" are "kaufen" and "verkaufen," respectively—close enough to cause confusion, as Tina Fey's character Liz Lemon discovers, to her chagrin, on an episode of the TV sitcom *30 Rock*.

Finally, and perhaps even more confusingly, some word pairs *seem* like they should have opposite meanings but do not. Let's consider the notorious case of "flammable" and "inflammable."[16] Many words in English that begin with the prefix "in" mean "not" something, as in "incorrect," "invisible," and "inarticulate." Therefore, it would be reasonable to conclude that "inflammable" means "not flammable."

However, the word comes from the Latin *inflammare* ("to inflame"). And as a term meaning "capable of being set on fire," it has been part of the English language since the early 17th century. In contrast, the first recorded use of "flammable" occurred more than two centuries later, in 1813. It was rarely used before the 1920s, but by the 1970s, its usage had surpassed that of "inflammable." What happened in this case?

Throughout time, it seems, people came to the realization that the confusion about "inflammable" was dangerous. Studies of fire safety terms repeatedly showed that many people equated "inflammable" with "nonflammable," which could lead to tragic consequences.[17] Starting in 1920, the National Fire Protection Association began to recommend that the word "inflammable" be replaced by "flammable"—a position echoed by the insurance industry.[18]

By mid-century, the issue was garnering high-profile attention. For example, a 1947 article in the *New York Times* noted that "some housewives . . . were under the impression that the word inflammable means non-inflammable and hence have been careless in handling certain canned or bottled fluids."[19] And by the late twentieth century, a concerted effort to deprecate "inflammable" had successfully relegated it to the backwaters of the language. The triumph of "flammable," even though it offends some purists, represents an unusual instance in which public safety was the engine for linguistic change.

DENOTATION AND CONNOTATION

One contributor to communication failure has to do with the meanings of words themselves. This may seem surprising since there is little debate about the referents for many, if not most, words. The meanings of terms like "table" and "chair," for example, seem fixed, unambiguous, and unlikely to cause misunderstandings. The use of some words, however, can engender controversy, negative reactions, and miscommunication. As a result, their use can create a source of ambiguity that counts as a strike against effective communication.

When thinking about the meanings of words, it's helpful to distinguish between their denotations and their connotations. The denotation of a given word is what one might find when consulting a dictionary: the accepted and conventional way or ways in which the term is used. Denotative meanings may change, but they typically evolve at a slow rate, and at any given point in time, there will be widespread agreement within a linguistic community about the referent for a particular term. The connotative meanings of words, on the other hand, are associations that a particular term might possess. These can be idiosyncratic and tied up in a complex and subjective net of both personal and cultural factors.[20]

In some cases, for example, a word may have connotations that can't be found in a reference book, or the term has usage issues that a dictionary definition only hints at. Euphemisms like "passed away" and "pushing up daisies" both refer to death, but only the first term would be appropriate when speaking to a grieving family member at a funeral. Whereas the former expression would be considered polite and appropriate in a funereal context, employing the latter would probably be perceived as rude or disrespectful. Lexicographers may attempt to provide some guidance about connotations by affixing usage labels, such as "informal" or "colloquial," to potentially problematic terms. But a dictionary can't anticipate the boundaries of acceptable usage in each and every situation.

In other cases, the connotative aspects of a word or phrase may change relatively quickly. Consider the term "social justice warrior." Its denotation is straightforward enough: it refers to someone who expresses or promotes politically progressive views. And in the early 2000s, its connotations were either neutral or positive. By the middle of the following decade, however, "social justice warrior" had become ideologically charged and was more likely to be used pejoratively. By the time it entered the *Oxford English Dictionary* database in 2015, it was labeled "informal, derogatory."[21]

Words or phrases can also sprout new meanings that are at odds with their original denotations. Let's consider "cowboy" and "Mickey Mouse" as examples. Originally, both terms had fixed and well-defined meanings: someone who herds cattle and the mascot of the Walt Disney Company, respectively. For most, they also engender positive connotations: romantic visions of the Wild West and a cartoon character beloved by children.

However, both terms have given rise to additional meanings that are distinctly different than their original denotations and connotations. "Cowboy" is now a label that is applied to someone who is brash or reckless or who threatens the use of force to get what they want. Several American presidents, such as Theodore Roosevelt, Ronald Reagan, and George W. Bush, have been criticized for engaging in "cowboy" diplomacy.

In a similar way, "Mickey Mouse" can be used as an adjective to refer to something that is amateurish or ineffective (a Mickey Mouse defense), too easy (a Mickey Mouse exam), or relatively worthless (a Mickey Mouse degree). These new meanings have become widespread and have been incorporated into dictionaries. As a result, the same term may be used by speakers in a positive or a negative way, and an ambiguous utterance like "He's a real cowboy" can lead to misinterpretation.

To complicate matters further, some words have a well-defined denotation, but controversy exists about whether it applies to a particular instance. A good example of this is the term "genocide," which would seem to have a specific and incontrovertible meaning: the deliberate and systematic slaughter of a large number of people who belong to a particular group. The Holocaust visited on European Jews by Nazi Germany serves as the paradigmatic example.

But what about the killing of as many as 1.5 million Armenians during World War I by the Ottoman Empire? More than a century later, the Turkish government refuses to characterize these deaths as an instance of genocide, although more than two dozen other countries have officially recognized it as such. Similarly, the "clearance operations" inflicted on the Rohingya by the government of Myanmar have been labeled as genocide by some UN agencies.[22] And how to differentiate the term "genocide" from related terms, such as "massacre" or "ethnic cleansing," is an open question.

Debate about how to characterize mass killings of the past, particularly the slaughter of indigenous peoples of the western hemisphere by colonial rulers, has gone on for decades. And in the early twenty-first century, the Truth and Reconciliation Commission of Canada concluded that the country had engaged in *cultural* genocide against the First Nations and Métis peoples through a policy of forced assimilation. Similar controversies have occurred about applying this term to other forms of persecution, such as the "reeducation camps" that the Chinese government has established for Muslim Uyghurs.[23]

Finally, it should be noted that some terms have a relatively neutral denotation but a strong negative connotation. "Diva," "courtier," and "scheme" are examples of words in which their negative connotations have become ascendent over their original, inoffensive meanings. The poster child for this phenomenon, however, would be the aversion that many Americans have to the word "moist." In the spring of 2012, *The New Yorker* ran a contest to determine which English word was most deserving of being voted off the linguistic island. The most frequently nominated term, by far, was "moist."[24]

In a study of aversive responses to this word, the psychologist Paul Thibodeau demonstrated that the issue doesn't seem to be the sound of the word—both "hoist" and "foist" are unobjectionable—but rather the disgust elicited by its association with bodily functions. Twenty percent of his participants demonstrated word aversion for "moist," and this negative association was stronger for participants who were younger, more educated, or female.[25]

As these examples suggest, one cannot always assume that there will be a precise match between the connotations of a word used by a particular speaker and her listener. In addition, nonnative speakers may lack the pragmatic awareness that native speakers take for granted and may mistakenly refer to someone as "taking a dirt nap" or "sleeping with the fishes" in the context of bereavement.

YOU CAN'T PLEASE EVERYONE

As we saw in the previous section, the connotations of words are powerful things. They are powerful enough that the negative associations of a single word played a role in ending the candidacy of a major presidential contender.

The candidate was Republican George Romney, and his résumé certainly made him a political force to be reckoned with. After a successful stint as chairman and president of the American Motors Corporation, he was elected to the governorship of Michigan and then reelected twice more. He was popular in the state and received 60 percent of the vote in 1966. And in mid-1967, he was polling nationally at 25 percent for the Republican nomination for the presidency. This put him in second place behind former Vice President Richard Nixon, the 1960 nominee.

The major issue in the 1968 election was the U.S. involvement in the war in Vietnam. The military escalation there had deeply divided the country. Romney had been an early supporter of America's involvement in the conflict and spent a month in South Vietnam in 1965. However, during a nationally televised interview in August 1967, he came out against the war, claiming that he had been "brainwashed" during his time in South Vietnam by the U.S. military and government officials.[26]

The use of this term recalled the brutal coercion endured by American prisoners of war by the Chinese army during the Korean conflict, causing some servicemen to cooperate with their captors or to defect. It was, therefore, an extreme and controversial term to use, particularly with regard to changing one's own mind.

Romney's statement was seized on by his opponents in both parties. By November, his poll numbers had dropped to 14 percent and continued to decline, but he soldiered on with his campaign. Bowing to the inevitable, he suspended his candidacy in late February. The following month, Nixon decisively won the New Hampshire primary on his way to his party's nomination and the presidency.

In certain cases, words with pejorative connotations have been rehabilitated. Examples would include "nerd" and "geek," terms that have been traditionally used to disparage socially awkward or eccentric individuals. However, as nerds and geeks like Bill Gates, Mark Zuckerberg, and Elon Musk became influential business leaders in technological fields, such terms were embraced by many as a badge of honor.

Another example would be "queer," which was used as a derogatory term for homosexuals as early as the 1890s. In recent years, it has been reclaimed by the LGBTQ+ community as a positive descriptor.[27] However, its use by straight men and women is problematic: does the speaker intend it in a positive or negative way? And even when straight people use it affirmatively, could this be considered a form of cultural appropriation? This is certainly true for the "n-word," which has been rehabilitated by Black Americans. However, its use by white speakers remains beyond the pale, so to speak.

Americans have long used the term "Canuck" to refer to Canadians, specifically French Canadians, in a derogatory way. In contrast, English-speaking Canadians use the term in a descriptive or even affectionate way. A minor-league Vancouver hockey team, established in 1945, adopted this moniker, and it was retained when a professional team was established there in 1970. And despite this clear embrace of the term by Canadians, many Americans remain uncomfortable about using it: its negative connotation lingers on like a bad odor in the minds of many.

Such phenomena are nothing new. Linguistic reappropriation occurred, for example, during the English Civil War, when the Roundheads (supporters of Parliament) referred to backers of King Charles I as "Cavaliers." Throughout time, however, the Royalists adopted this epithet to refer to themselves.

More than a century later, the term "Yankee" was used by the British as an unflattering appellation for American colonists. Its origins are unclear, but it may originally have been a derogatory term applied by the British to the Dutch settlers in the New World (essentially, a corruption of the Dutch term for "Johnny").[28] The Brits used the term to stereotype colonial men as low class and unsophisticated (as in "Yankee doodle dandy").[29] And a century after that, "Yankee" became a derisive term used by the Confederates to refer to their Union adversaries.

"Yankee" still functions as a negative epithet in the South to refer to those living in the North. However, it also occurs in positive contexts as well, as in the phrase "Yankee ingenuity." And fans of the Bronx Bombers don't perceive it as a negative term.

Finally, there are terms for which even the denotative meaning is ambiguous. An example would be "bimonthly," which can be used to refer to something happening once every two months or twice a month. Some guides, like *The New York Times Manual of Style and Usage*, insist that "bimonthly" means every two months and that "semimonthly" means twice a month.[30]

Other usage guides are content to wave the white flag of surrender: some terms are simply too problematic to be used and should be avoided. Bryan Garner, for example, suggests that "for the sake of clarity, you might do well to avoid the prefix [bi-] when possible."[31] And in a usage note for "bimonthly" and "biweekly" on their website, the editors of the Merriam-Webster dictionary appear to have thrown up their hands about which usage is correct: "This ambiguity has been in existence for nearly a century and a half and cannot be eliminated by the dictionary."[32]

When the denotations of a word are contradictory or when its connotations change throughout time, there may be a general tendency to avoid using that term. This phenomenon was given a name by Bryan Garner in his *Modern American Usage*: he refers to such terms as having been "skunked."[33] In his discussion of the phenomenon, he mostly cited words whose meanings have broadened throughout time, such as "decimate," "fulsome," and "hopefully." Each of these terms had more specific meanings at an earlier point in time, but they are now usually employed less precisely.

"Decimate," for example, originally meant "to kill one out of 10." It's now typically used to refer to the wholesale destruction or vanquishing of one's opponent. Garner's argument is that as a word undergoes such a transition, it becomes impossible to please everyone: sticklers will object to the word being used in a new way, while the linguistically progressive will perceive those clinging to the traditional meaning as out of step with the times.

Garner's logic also applies to the words described earlier. It is easy to imagine an American avoiding using the word "Canuck" to describe his friend from Montréal or the straight reporter who decides not to use the term "queer" in an article about a pride rally. Connotations are powerful and tricky things and provide fertile ground for misunderstandings and miscommunication to take root.

YOU JUST HAD TO BE THERE

> *A man standing on a riverbank shouts to a man on the other side, "Hey, how do I get to the other side of the river?"*
>
> *The second man responds, "You* are *on the other side."*
>
> —r/dadjokes, Reddit, 2017

If you're in the habit of eavesdropping on other people's conversations in restaurants or in other public spaces, you've probably noticed that some of the things you overhear are difficult to understand. Consider the case of a woman asking her dining companion, "Is this as good as it was the last time we were here?"

As an interloper, you would have no way of identifying the subject of this question. Is she referring to the food? The service? Their interaction? And when was "the last time"? Was it last week? Last month? Ten years ago? Similarly, who is "we"? It's likely that the woman asking the question means herself and her dining companion. However, the range of possibilities is vast: it could, for example, refer to both of them as well as to the royal family of Monaco, with whom they happened to be dining on the previous occasion.

Even the word "here" is problematic: it could mean the restaurant that all of you are currently in, or it could be a reference to the neighborhood, city, state, or even country. But if the dining companion recalls the previous episode—and by asking the question in this way, the speaker is assuming that he will—there is no need for the question to include such details. This is another example of how shared common ground, discussed in chapter 2, works its referential magic.

In most cases, the referent for words and phrases is clear and unambiguous: "the sun" is that bright thing up in the sky, and "my dog" is the loyal creature asleep at my feet. The problem in the restaurant example is that the referents for certain words depend crucially on who the speaker is or when and where the conversation is occurring.

The word "I" is a good example. It points to the person who is currently speaking, and its referent changes depending on who is doing the talking. "Now" refers to a precise instant in time. "Here" and "there" are spatial terms that are relative as opposed to absolute. Julius Caesar could refer to "yesterday" and "today" (well, *hesterno* or *hodie*, anyway), and you can as well, although the dates will be a bit different.

In languages like English, there are a number of *deictic* terms (from the Greek for "to show") that refer to people, space, and time. Deictic terms can't be defined in an absolute way, and as a result, their use can create ambiguity and confusion, as in the joke at the beginning of this section.

In other cases, a word can be problematic because it has more than one possible referent. Consider the following:

> *He took a swig from the bottle of beer that had*
> *been placed before him on the bar.*
> *It was warm.*

It may not be completely clear what the pronoun "it" refers to in this case. However, most people would conclude that it's the beer that was warm since that is the object of the first sentence. A moment's thought, however, will reveal other potential candidates for the identity of "it." Some of these rely on inferences based on world knowledge. For example, people drink beer when the weather is warm, so "it" could refer to something that wasn't even mentioned in the earlier sentence: the weather. Finally, it's even possible that the referent is to the bar; like the beer, it was also mentioned explicitly. The problem with this interpretation, however, is that we don't often notice or remark on the temperature of surfaces in our environment. As a result, our world knowledge could be used to conclude that this interpretation is possible but also unlikely.

Linguists refer to cases like these as instances of *anaphora* (Greek for "carrying back"). And as the example with "it was warm" illustrates, figuring out the correct referent isn't always a straightforward enterprise. The process by which ambiguous words are matched to their referents—referred to as anaphor resolution—is a major topic of interest in linguistics and computer science.[34,35]

As researchers attempt to create programs that can understand human language, they must find ways to mimic the inferential processes that are easy and seemingly effortless for people. And since these programs lack knowledge of the world and what people typically talk about, this can be tricky.

The ambiguity created by anaphors can also be a source of humor, as in the following sentence from a British air raid leaflet from World War II:

> *If an incendiary bomb drops next to you, don't lose your head. Put it in a bucket and cover it with sand.*[36]

Once again, world knowledge allows us to infer that the "it" refers to a bomb and not your head since we don't typically place our body parts in buckets. This inferential process feels effortless and automatic, and the mental image conjured up by considering the alternative is amusing. A computer program with no knowledge of incendiary devices or anatomy, however, might find assigning the referent to be more challenging.

Let's consider a final case that might be more problematic:

> *Mary saw Jane while she was walking down the street.*

In this sentence, "she" could refer to Mary or to Jane. Mary is mentioned first, however, and this can serve as a signal that any anaphoric expression will refer to her.[37] However, speakers of English also tend to resolve such ambiguities by assuming that "she" refers to the subject of the sentence (in this case, Mary) instead of the object (Jane).

This can create confusion for those who learn English as a second language since in languages like French, anaphor resolution favors the object of the sentence.[38] As a result, miscommunication can arise from factors as basic as figuring out who is doing what to whom.

JARRING JARGON

Jargon is an almost inevitable consequence of the need to make use of specialized terminology within a given profession or occupation. And although a shared jargon can function as a unifying force within a group of practitioners, such terms can create confusion for the laypeople who encounter them.

There is an ongoing debate within the medical community, for example, about whether patients should be allowed to see their physician's notes. Good arguments can be made both for and against such a policy. A recurring source of concern, however, is whether patients would be able to understand the technical terms, abbreviations, and jargon that are commonly employed by doctors and other medical personnel.

For example, physicians frequently use the acronym SOB to refer to "short of breath," but it's easy to imagine someone reading "The patient appears SOB" and taking umbrage with such a description. Similarly, a notation of NERD, which stands for "no evidence of recurrent disease," might be misinterpreted as well.[39]

The legal profession is notorious for its use of obscure phrases and opaque terminology. This is partly due to an accretion of vocabulary throughout time from such languages as Latin (amicus curiae, habeas corpus, pro se) and Anglo-Norman, which dates from the conquest of England nearly 1,000 years ago (estoppel, force majeure, voir dire). And even when English is employed, jurors and other neophytes in the courtroom may have to contend with concepts like "attractive nuisance," "fruit of the poisonous tree," or "peremptory challenge."

The military is famous for its extensive use of acronyms, some of which have escaped the confines of the armed forces and have entered common usage. AWOL (absent without leave), first used in the late nineteenth century, has made the transition, as has SNAFU (situation normal: all f——ed up), which made its appearance during World War II. The various military branches also employ an extensive and colorful set of jargon terms. Some of these are inferable: "brain bucket" (helmet), "chest candy" (ribbons and medals), or "meat wagon" (ambulance). Others, however, are relatively opaque ("farts and darts," "ring knocker," "shavetail").

Aficionados of certain pursuits, such as wine tasting, have developed extensive and highly insular vocabularies that are largely impenetrable to outsiders. Descriptions like "a dense, smoky, medium-bodied wine with a good finish" can truly only be understood by others who have invested the time and energy to educate their palates.

Skiers and snowboarders have many terms to describe the relative degrees of compactedness and iciness of snow ("powder," "crud," "crust"). Surfers have many ways to describe waves. Heraldic terminology, which involves descriptions of the parts of coats of arms, also requires a great deal of specialized jargon. Model railroaders have their own specialized argot. In short, mastering virtually any profession or avocation involves acquiring a lingo that only those in the know can appreciate and understand.

In the context of miscommunication and misunderstanding, the use of jargon raises some interesting issues. When directed at laypeople, such specialized terminology is not typically intended to be opaque. It seems likely that practitioners simply forget or don't fully realize that they're interacting with someone who isn't familiar with concepts and terms that the expert takes for granted. In such cases, the issue is best thought of as a failure of common-ground monitoring. In other situations, the use of jargon may be intentional. It can be used as a way of determining whether someone is truly part of a particular profession or is knowledgeable about a certain domain.

Concerns about the dangers of legal jargon are voiced on a regular basis. A 1977 article in the *New York Times*, for example, lamented that legalese had become such a problem in the profession that there were concerns about whether attorneys and judges could even understand each other. It cited a poll in which a majority of judges and lawyers found the opinions rendered by the U.S. Supreme Court to be "too long and confusing."[40] In 1984, law professor Robert Benson called for an end to legalese, asserting that "the game is over."[41] Despite such concerns, a 2017 article by Martin Schwartz bemoaned the unfettered use of jargon in legal documents like real estate transfers.[42]

It should be noted, however, that legal language does have its defenders. Writing in the same year as Schwartz, Michael Stephenson pushed back against the so-called Plain Language Movement, arguing that a wholesale abandonment of legal terminology is "not the panacea for every perceived ill of legal language."[43] Legal documents, such as contracts, need to address all possible contingencies, and doing so requires a degree of precision that simpler language simply cannot provide. From this perspective, precise legal terminology is essential to prevent misunderstandings.[44]

Similar concerns about jargon have bedeviled the medical profession. A 2011 poll found that only 8 percent of the public truly understood what practitioners meant when they used the term "palliative," and even references to more common concepts, such as "tumor," were problematic.[45] And as with legalese, there is concern that not all fellow medical practitioners understand health-related terms.

In 2006, a British journal read by general practitioners reported that common ophthalmologic jargon was frequently misunderstood by medical generalists. A survey revealed, for example, that only one-sixth of primary care providers were able to correctly decipher the abbreviation "PVD" (posterior vitreous detachment). A majority of the general practitioners incorrectly responded that it stood for "peripheral vascular disease." Overall, two-thirds of the respondents were unable to explain or decipher 12 common ophthalmologic abbreviations.[46]

Jargon can't be eliminated by fiat or through a naive belief that people will voluntarily educate themselves in arcane terminology. The widespread availability of English dictionaries during the past 400 years, for example, doesn't seem to have reduced concerns about the pernicious effects of jargon. And increasing specialization within professions suggests that this problem will only intensify in the future. Jargon seems destined to live on and to be as useful within a discipline as it is mystifying to the uninitiated.

5

WORDS, PART 2

In addition to the problematic word types described in the previous chapter, miscommunication can also arise when people make use of euphemisms, idioms, and metaphors. Metaphors, for example, are helpful in comparing things that are similar, but since such mappings are never exact, their use can be problematic. People also commonly employ double entendre and innuendo, but sometimes these references are oblique or too subtle. As a result, they are not understood. And to make matters even more confusing, a word can suddenly enter the language or shift rapidly in its meaning or acceptability. Let's consider some concrete examples.

WHAT'S THE GOOD WORD?

Early in the afternoon of November 3, 2014, the devotees of Chuck Grassley's Twitter account were treated to the following observation:

Windsor Heights Dairy Queen is good place for u kno what

For a day or two, pundits on the internet were abuzz about his cryptic pronouncement. Within a few days, it had been retweeted more than 2,700 times and liked more than 3,200 times.[1]

Was Grassley implying that the central Iowa establishment was a prime spot for carnal or illicit activity? Were there pleasures to be had there other than soft-serve ice cream?

"You know what" is certainly an expression that can refer to sexual activity. To make sense of Grassley's tweet, however, it may be helpful to know that, at the time he composed it, Grassley was the 81-year-old senior senator from the state of Iowa. And as a father of five, he was certainly familiar with at least one meaning of "you know what." But those who followed Grassley's Twitter account also knew him to be an inveterate and enthusiastic user of the social media platform. He often posted about unusual events, such as hitting a deer on the highway, or even mundane ones, like burning piles of brush.[2] And his followers were also familiar with his proclivity to employ creative spellings and shortenings in his posts.

Grassley himself, when asked about his intent by a Des Moines reporter, replied, "I wanted to give Windsor Heights Dairy Queen some credit for making good Dairy Queen and doing you know what. And what do you do at Dairy Queen, you eat Dairy Queen." He elaborated that on that particular occasion, he had indulged in vanilla and chocolate ice cream.[3]

Euphemisms (Greek for "to speak fair") are often employed to avoid more direct references to taboo subjects, such as sex, or unpleasant ones, such as death. And although the confusion created by Grassley's tweet is amusing, there are also important real-world implications associated with the use of euphemism. This is clearly an issue for the medical community since doctors must often refer to serious and potentially life-ending conditions when talking to their patients.

A study of British physicians, for example, examined their use of the term "heart failure" as well as euphemisms that they commonly employ to refer to this condition. Two frequently used substitutes are "You have fluid on your lungs, as your heart is not pumping hard enough" and "Your heart is a bit weaker than it used to be."

The authors of the study point out the doctors often find themselves in a double bind when talking to their patients. To avoid alarming their charges, the physicians reported making extensive use of such circumlocutions. At the same time, however, these euphemisms may cause patients to believe that their condition is less serious than it really is, which may lead to lower rates of compliance with their physicians' recommendations.[4] As it turns out, even euphemisms for death itself, such as "gone over to the other side," can create confusion.[5]

It is also the case that such euphemisms vary in their delicacy or appropriateness. For example, references to "your loss" or to the deceased as "being in a better place" would be appropriate circumlocutions when consoling a grieving family member at a funeral home. However, as was discussed in chapter 4, the use of other idioms to refer to death, such as "pushing up daisies" or "taking a dirt nap," would not be. It is all too easy to imagine how a nonnative speaker of English, seeking to avoid a direct reference to death, might mistakenly employ one of these less appropriate expressions.

The specter of death and a recourse to euphemism is to be found not only in the hospital or the funeral parlor. The military also makes extensive use of such circumlocutions to downplay the lethality of its operations. As William Astore has argued, terms such as "collateral damage" (civilian deaths), "enhanced interrogation" (torture), and "extraordinary rendition" (kidnapping) have been devised and employed by political and military leaders to make such activities more palatable to the public.[6] "Friendly fire" can be anything but congenial, a "Department of Defense" can initiate offensive action, and "kinetic" warfare is a more genteel way of referring to the physical violence of weapons that can maim or kill.

Such coinages are nothing new, however; the historical record is replete with such euphemistic terms and slogans. A good example would be the Japanese promotion of a "Greater East Asia Co-Prosperity Sphere" during World War II. This phrase was used by Imperial Japan to refer to the beneficent replacement of the colonial powers that had previously controlled Southeast Asia. Instead of bringing prosperity to these "liberated" territories, however, the Japanese occupation brought untold misery to millions who endured economic exploitation or were conscripted as slave laborers.[7]

Finally, some euphemisms are simply ineffective. Morton Gernsbacher and her collaborators explored the use of the term "special needs" as a way of referring to people with physical or mental disabilities. If the purpose of a euphemism is to replace one term with a kinder and gentler alternative, then "special needs" seems to have failed in this regard. The researchers found that people described as having special needs were perceived more negatively than when they were characterized as having a "disability" or as having a specific kind of handicap, such as being blind or having Down syndrome.[8]

It seems clear, then, that the laudable goal of cloaking certain topics in more positive language can easily go astray and can create miscommunication when referring to medical conditions, death—or even ice cream.

IF YOU KNOW WHAT I MEAN

A woman walks into a bar and orders a double entendre,
so the bartender gives it to her.

—Unknown wit

Although double entendre is closely associated to forms of verbal wordplay such as puns and innuendo, such utterances can be used for a wide variety of purposes, such as to be flirtatious or to drop a hint. They can even be used to deliver insults and threats.

Many perfectly innocent English words, such as "climax," "screw," and "hard," have secondary meanings that are risqué in nature. But there are plenty of non-carnal examples too. At the conclusion of the film *The Silence of the Lambs*, Hannibal Lecter calls Clarice Starling and says, "I do wish we could chat longer, but . . . I'm having an old friend for dinner." Conventionally, the word "having" would mean "dining with," but since Lecter is a cannibalistic serial killer, he may also be implying that a more sinister culinary experience is in the offing.

Many authors and playwrights, such as Homer, Chaucer, and Shakespeare, are well known for their sly use of double entendre. Certain actors, such as Mae West and Groucho Marx, are also strongly associated with the form. Double entendre as an art form is alive and well in the twenty-first century, as in the lyrics of hip-hop artists.

The Bible might not spring to mind as a book containing sexual innuendo, but a number of scholars have suggested otherwise. In the second chapter of Joshua, for example, the leader of the Israelites sends two spies to reconnoiter in Jericho. They arrive at the house of a woman named Rahab, where they spend the night. The text describes Rahab as a prostitute and states that the men "lodged" in her home. However, the original Hebrew wording eschews the usual term for lodging and instead makes use of an expression that also means "lay down," strongly implying that the men had sex with Rahab.[9]

This double entendre is lost in English: the King James Version simply states that the men "lodged there," while the New International Version relates that the spies "stayed there." Because of its dependence on the same word having multiple meanings, double entendre often does not survive in translation.

In some cases, scholars are divided on whether a writer intended to commit a double entendre. A good example of this appears in Dickens's *Oliver Twist*. One of Fagin's young pickpockets is named Charley Bates, and this character is mentioned nearly 100 times in total. In about half of these references, he is Charley Bates or Master Charles Bates, but in the other half, he is "Master Bates."

Since other male characters in Dickens's novel are not referred to as "Master," some critics have argued that the double entendre was intentional. Many of Dickens's characters do possess odd or humorous names, although this one comes across as broad and somewhat puerile, at least to twenty-first-century readers. Other critics, however, have claimed that "masturbate" was not a term commonly used in Dickens's day.[10]

In other instances, the "double" aspect of a double entendre has been lost due to shifts that occur within a language. This is the case for some of Shakespeare's innuendo. A frequently cited example is Hamlet's exhortation to Ophelia in act 3, scene 1. He encourages her to get to a "nunnery," a command that he repeats five times—it's hardly meant as an offhand remark.

In one interpretation, a jealous Hamlet seems to be entreating the woman he loves to shut herself away from all other men. However, he's also angry with her for ending their relationship, and this is reflected in the other meaning of "nunnery" in Shakespeare's day: it could refer to a brothel as well as to a convent.[11]

It's worth noting that the process can also work in reverse: words can take on new and distinctly different meanings that they did not originally possess. For centuries, the word "gay" had no sexual connotation: it meant lighthearted, cheerful, and carefree. Poetry was once known as "the gay science." Throughout time, however, the word also came to mean uninhibited and flamboyant. And by the 1920s, it had become a slang term for homosexuality. Children might snigger at phrases like the donning of "gay apparel" in "Deck the Halls" (the carol dates from 1862) or "We'll have a gay old time," an assertion that appears in *The Flintstones* theme song in the early 1960s. Clearly, words and phrases can take on double meanings as well as shed them.

Miscommunication can occur when the person on the receiving end of a double entendre understands only its overt meaning and fails to perceive a second interpretation. The speaker may resort to a wink or a nudge or a phrase like "If you know what I mean" to underscore the intended witticism. The reverse can happen as well: a listener may call attention to an innocent remark that could have another meaning by saying something like "That's what she said." This rejoinder was popularized by the NBC sitcom *The Office*.[12] In this way, a statement can be effectively transformed into a double entendre.

Whether the original speaker appreciates his partner's contribution, however, is another matter. If someone is attempting to have a serious conversation and their partner interrupts to point out an alternative and juvenile interpretation, the speaker is likely to be annoyed. However, if the allusion sparks amusement, both parties may attempt to extend the wordplay as the conversation continues.[13]

In theory, triple entendre are possible, but it can be surprisingly difficult to craft an utterance in which three distinct meanings, all relevant to a particular context, are invoked by the same phrase. One example occurs in *Hairspray*, the 1988 film by John Waters. At the end of the energetic number "It's Hairspray," a paean to the eponymous styling product, Corny Collins remarks to a female dancer, "Hey baby, you look like you could use a stiff one!" In this case, "a stiff one" could refer to a rigid coiffure, a strong drink, or the ministrations of Collins himself. The dancer, in turn, makes a theatrical wink toward the audience to underscore the more suggestive meaning of the remark.

Another example can be found in the musical *Hamilton*. In "My Shot," the phrase "I am not throwing away my shot" refers to alcohol being drunk by the cast and also to Hamilton's desire for fame. In addition, it foreshadows Hamilton's duel with Burr, in which he will deliberately miss his opponent.

WHAT'S A METAPHOR GOOD FOR?

During the early years of the twentieth century, physicists debated the best way to characterize the structure of the atom. Researchers had determined that atoms were composed of smaller particles, but how were they arranged?

In 1904, the British physicist J. J. Thomson proposed that negatively charged electrons are uniformly distributed throughout a positively charged atomic medium, where they are embedded much like plums in a pudding. This "plum pudding" model, however, was challenged by subsequent experimental findings. To account for these results, Thomson's student Ernest Rutherford proposed an alternative. He hypothesized that most of the mass, as well as the positive charge of an atom, exists in a small space at its center.

The Rutherford model of the atom characterizes it as having a nucleus surrounded by orbiting electrons. His description is often referred to as the planetary model, based on its similarity to the solar system. According to this way of thinking, the atomic nucleus is like the sun, and the electrons are like the retinue of planets moving around the central body.[14]

It's important to realize that these physicists were making use of analogy and metaphor in trying to characterize atomic structure. That is, they were attempting to explain something new by making use of concepts that were already familiar to other scientists, such as the notion of a nucleus (from cellular biology), orbits (from celestial mechanics), and even plums (botany).

When most people think of metaphors, they are reminded of their use as literary devices in poetry or perhaps in the plays of Shakespeare. By comparing Juliet to the sun, for example, Romeo means that his beloved is a source of energy, warmth, and comfort. Poets and playwrights find metaphors to be a productive way to express their ideas, but they are essential for science as well.

Metaphors and analogies are powerful tools for illuminating new concepts or for thinking about things in new ways. However, they also have important limitations. No two things are exactly alike, and therefore any metaphorical comparison will inevitably be imprecise. Some attributes will be shared by the concepts being compared, but many will not be.

Because Juliet is a human being, for example, we automatically filter out many irrelevant attributes that are implicit in such a comparison. We don't conclude that Romeo thinks she is a ball of hydrogen and helium, that she emits vast numbers of photons, or that she is located millions of miles from Earth. And part of the pleasure of reading literary works is working out and appreciating those mappings that are relevant.

The limitations of metaphor come to the fore when we consider how educators explain complex concepts, like the structure of the atom, to their students. Rutherford's planetary model, while an improvement on plums in pudding, still has a number of shortcomings.

These deficiencies caused Rutherford and Niels Bohr to revise the model, principally to take into account quantum phenomena. And the Bohr model itself has been refined and made more complex by later generations of theorists. How should teachers explain all of this to their students? The simple answer is that they cannot and do not.

Secondary school instructors typically expose their pupils to a hybrid model of the atom, drawing on Rutherford's and Bohr's notion of a nucleus and electron orbits. Such an approach is simplified by design—it omits the notion of atomic orbitals or "clouds" of electron densities, for example—but is designed to expose students to the basic rudiments of atomic structure.[15]

Many of these schoolchildren, however, will never have their understanding of atomic structure refined by more advanced coursework, and as a result, the notion that an atom is like a tiny solar system is entrenched in the public's understanding of physics. Even the seal of the Atomic Energy Commission perpetuates this oversimplification, depicting a nucleus with four electrons traveling in orderly orbits around it.

Lurking at the heart of metaphorical comparisons is an even more insidious issue and one that is frequently overlooked. Metaphorical mappings involve a subject, which is called the topic, and the thing being compared, known as the vehicle. When Romeo says that Juliet is the sun, Juliet is the topic, and the sun is the vehicle.

To understand this mapping, the attributes of the vehicle must be well known. And as Keith Taber has pointed out, this is the other danger in comparing the atom to the solar system: it assumes that students are *already familiar* with the structure of the solar system. And an imperfect understanding of the vehicle will invariably result in an imperfect mapping onto the target.

An understanding of the structure of atoms is but one example of a far larger issue. Some language researchers have argued that metaphors function as lenses through which we view and make sense of the world around us. So-called conceptual metaphors, like "argument is war" or "love is a journey," allow us to understand abstract psychological states in terms of more tangible concepts.[16]

The linguist Michael Reddy, for example, has described how humans spontaneously make use of a "conduit" metaphor to describe how ideas are transferred from one person to another. We make use of a number of core metaphorical expressions to describe this process, such as *putting* our thoughts into words or *giving* someone an idea.[17] Even though such statements are inherently metaphorical, we usually don't consider them as such, which limits our ability to conceptualize communication in some other way.

Metaphors, then, are double-edged swords, highlighting important similarities while obscuring essential differences.[18] Within a given culture, metaphors can create miscommunication, and even those that appear to have cross-cultural similarities may still conceal important differences.[19] Once again, we see how a certain type of language can create ambiguity and count as a strike against effective communication. Metaphors are good for many things, but they also contain within them the seeds for misunderstanding.

IDIOMATICALLY SPEAKING

Orrin Hatch was irritated. On August 7, 2017, the Senate's Finance Committee chairman was fed up with changes being proposed for the nation's health care laws. Congress was preparing to adjourn for a monthlong recess, leaving him little time to draft legislation for tax reform. Giving voice to his frustration, Hatch told *Politico*, "We're not going back to health care. We're in tax now. As far as I'm concerned, they shot their wad on health care and that's the way it is. I'm sick of it."[20]

Many commentators were surprised that the 83-year-old senator from Utah would use a crude sexual term to describe the proposal to make changes to Obamacare. Taken aback by this reaction, Hatch's office fired off a tweet intended to provide a "jargon lesson on 'wads' and the shooting of them." The tweet included a screen capture from the *Oxford English Dictionary* that defined the phrase as meaning "to do all that one can do," as well as the use of "wad" to refer to a plug to keep the powder and shot in position in the barrel of a gun.[21] (It's worth noting that this entry in the dictionary hadn't been updated since 1921 and does not include a definition that refers to ejaculation.)

Clearly, idiomatic expressions have qualities that make them fertile ground for miscommunication. Other figurative expressions, such as metaphors, can be identified by their clear semantic violations: William Shakespeare's "All the world's a stage" or Pat Benatar's "Love is a battle-field" are clearly nonliteral comparisons. Idioms can't be picked out in the same way.

Consider the expression "To open a can of worms," which means to consider a complicated situation that may cause further problems. This has a different feel than comparing the theater to daily life: those concepts do have a conceptual overlap, as do affairs of the heart and human conflict. It would be difficult, however, to find an obvious connection between prepackaged invertebrates and complex situations.

Language researchers refer to such seemingly arbitrary expressions as being "opaque," and they are the bane of second-language learners around the world. Idiomatic expressions are common in all languages, but their existence can't be predicted: a particular tongue might have an idiom for making small talk, for example, but then again, it might not. Some idioms function as euphemisms: as we've seen, they're used in polite company to avoid talking about unpleasant topics like death.

Other idioms have more in common with slang and would not be used in any sort of formal setting: consider the difference between being described as "in a family way" or being "knocked up." And someone can know and understand all the words in such an expression but be no closer to comprehending it. This characteristic of idioms means that they can't be decomposed; in other words, their meaning can't be divined by analyzing the constituent words that make them up.[22]

These issues aren't only a problem for nonnative speakers: many idiomatic expressions are regional, while others quickly fall out of linguistic fashion. And some, like the one employed by Senator Hatch, have seen significant shifts in meaning throughout time. World knowledge also plays a role. Some idioms require familiarity with popular culture or sports that may not be well known throughout the entire English-speaking world.

British English has many such expressions that are opaque to Americans. A good example is "to do a Devon Loch." Understanding this idiom, which means "to fail suddenly," requires knowing about the collapse of a racehorse by that name in the 1956 Grand National. Most Americans know that "a sticky wicket" refers to a difficult situation (a can of worms, if you will), but it can't be fully understood without some knowledge of the role of wickets in the game of cricket.

As we saw in the example with Senator Hatch, idioms can vary widely with regard to their familiarity. The author remembers the confusion he once sowed in the mind of a graduate student who was a generation younger than he. I told her that I wanted to put a bug in her ear. I used the expression as a way of saying that I wanted to make a suggestion for her to consider. I distinctly remember how her eyes widened and the suspicious look that followed. At the time, I thought that she was reacting negatively to my recommendation about her research project. Much later, she admitted to being unfamiliar with the phrase, although she had been able to figure it out from the context of our conversation. Her initial reaction to the literal meaning of the phrase, however, had been one of shock and disgust.

In addition, idioms are problematic because they are frozen: the words in such expressions can't be altered without breaking the idiomatic connection. The flowers in the death idiom "pushing up daisies" must be daisies and not marigolds or sunflowers. Wishing someone success by telling them to break both their legs is not twice as good as telling them to "break a leg": it's merely odd—and sounds rather mean spirited.

Metaphors, on the other hand, are much more productive and flexible in this regard. A few years before Pat Benatar informed us that love is a battlefield, Bette Midler had opined that love is a river, a razor, a hunger, or a flower.

But are idioms as opaque as they seem to be? Let's open a can of worms by reconsidering "to open a can of worms." The *Oxford English Dictionary* attests that the phrase appeared in written English as early as 1962. It doesn't offer an etymology, but the website *Mental Floss* reports that in the mid-twentieth century, bait shops sold earthworms in sealed metal cans with handles and lids.[23] Opening the container meant that the live bait could crawl out, causing problems in much the same way as opening Pandora's box, which is a much older idiom. Although this may be nothing more than a folk etymology, the phrase becomes more transparent when viewed in this way.

To take this one step further, it has been suggested that many seemingly arbitrary expressions are manifestations of underlying conceptual metaphors. George Lakoff and Mark Johnson, in their book *Metaphors We Live By*, proposed that such metaphorical frameworks exist for many phrases that otherwise appear to be unrelated.[24] As an example, consider the conceptual metaphor "love is a journey." It can be seen as the latent connection that exists behind the disparate phrases that people use to describe their relationships. Examples would include "we're off to a great start," "we're spinning our wheels," "we're stuck," and "we're at a crossroads."

And the psychologist Ray Gibbs has proposed that the conceptual metaphor "anger is heated fluid in a container" provides a framework that unites idioms like "blow your stack," "flip your lid," and "hit the ceiling."[25] Conceptual metaphors may not account for all idiomatic expressions, but it may be the case that some are less arbitrary than they at first appear.

THE (LINGUISTIC) TIMES ARE A-CHANGIN'

Many words and expressions fall out of usage because of changes in the culture or because of concerns about a term's origins or original meaning. Members of older generations, who learned these expressions decades earlier and may still use them, can find themselves at odds with members of younger generations, whose sensibilities may differ significantly from their own.

A good example of this would be the expression "young Turks," a term that has been used in English since the mid-twentieth century to refer to younger people who are eager to make sweeping changes, radically innovate, or otherwise shake up the established order within an organization. Until very recently, it had a neutral or even positive connotation.

Throughout time, however, the origins of the term have become more commonly known. It involves the 1908 overthrow of the Ottoman Empire's absolute monarchy and the subsequent coup by one such group, led by the Three Pashas, in 1913. They were responsible for the Ottoman Empire's alliance with the Central powers in World War I and the Armenian genocide that occurred during and immediately after that war. As awareness of this atrocity has spread, the phrase began to be seen as problematic by many. A style guide published in 2005, for example, states that "young Turk" is no longer acceptable and that its use should be avoided.[26]

Some terms are clearly problematic, and there is little disagreement that their use should be avoided. Words used in the early twentieth century to refer to those with low intellectual ability, such as "idiot," "imbecile," and "moron," are clearly verboten. A related term, "retarded," has had a more complex history. It was originally thought of as a more benevolent term until it too became seen as an insensitive descriptor. The same is true for "mentally challenged," which has been displaced by the more contemporary "special needs" or "exceptional."

The psychologist Steven Pinker refers to such developments as the "euphemism treadmill": any such term will become pejorative throughout time, leading to the adoption of a new term, which will become problematic in turn.[27] A similar historical progression can be seen with the terms "shell shock," "combat fatigue," "operational exhaustion," and "posttraumatic stress disorder."

The problem with such terms is that the perception that they are problematic or objectionable may move relatively slowly through a linguistic community. Typically, younger people are more attuned to such terms falling out of fashion, and they may be offended when older adults use them without any awareness that they have become tainted or unacceptable. And these issues are tied up in larger debates about political correctness and political ideologies.

In general, terms that are associated with stereotypes of racial or ethnic groups have fallen out of favor and out of use. These include not only the names of groups, such as Eskimos (Inuit) or Indians (Native Americans or First Peoples), but also actions, such as "going off the reservation" (behaving independently), "shanghai" (abduct), or "gyp" (swindle).

Dictionaries typically provide little guidance about using such terms. This is because many of them are descriptive as opposed to prescriptive; that is, they provide a record of how a word or phrase has been used, not whether it *should* be used. In earlier times, the makers of dictionaries often provided guidance about usage, employing labels such as "disparaging," "offensive," "obscene," or "vulgar." However, it takes many years to revise a dictionary, and even then, lexicographers may be reluctant to address the complex and subjective aspects of shifting sensibilities.

One problem with such shifts is being aware that they have occurred, while another is determining the line between what is acceptable and unacceptable. An example of this would be the use of the term "Oriental" to refer to the inhabitants of the countries in the Far East and the Pacific. Its use was banned in state documents in New York in 2009, and President Obama signed legislation that eliminated the word "Oriental" (as well as "Negro") from federal laws in 2016. But even some Asian Americans were surprised by this action since the term had never been widely perceived as a racial epithet.[28] And the word lives on in the language as an adjective, as in Oriental languages, Oriental rugs, and Oriental topaz.

In some cases, a term has been unfairly maligned due to the rise of a folk etymology. A good example of this is the phrase "rule of thumb," which was originally employed in the seventeenth century to refer to a rough measurement. In contemporary usage, it refers to a practical method or a heuristic. However, many people believe that the term derives from the practice of allowing husbands to beat their wives with a stick—as long as the rod wasn't thicker than the man's thumb.

Supposedly, this rule dates back to a ruling by a British magistrate, Francis Buller, in 1782. However, no record of such a decision exists. Its connection to wife beating seems to date from the 1970s, and awareness of this folk etymology has spread widely since that time.[29] In recent years, many people have avoided using the term because of this mistaken association with domestic violence.[30]

On the other hand, a term with a truly problematic etymology may be commonly employed because its history is not well known. An illustration of this can be found in the phrase "peanut gallery," which refers to a noisy or disorderly group of spectators in an audience. For many older adults, this term is redolent with fond memories of *Howdy Doody*, the NBC show for children that was popular in the 1950s. On that program, the term referred to the onstage bleachers for kids, who sang "It's Howdy Doody Time" at the beginning of each episode.

The phrase, however, dates back to late nineteenth-century vaudeville performances and refers to the cheapest seats at a venue. The customers seated in this area of a theater would not hesitate to throw concession snacks at performers who displeased them, and this resulted in untalented (or unlucky) entertainers being pelted with peanuts from these patrons.

These inexpensive seats, typically high up in the balcony, were frequently reserved for Black patrons, and therefore some have argued that this is a racist term and should be avoided.[31] Lexicographers, however, have been unable to definitively determine whether the expression is racially motivated or whether it is more generic, referring to those of lower social classes.

It should be clear from these examples that misunderstandings can easily arise when people have diverging views about the acceptability of a particular expression. And since the writers of dictionaries have become less comfortable with playing the role of umpire, such disputes tend to be waged in the court of public opinion.

WHO DECIDES WHAT WORDS MEAN?

In the 2004 film *Mean Girls*, a character named Gretchen, played by Lacy Chabert, repeatedly uses the phrase "So fetch!" to express her approval. After employing it to comment favorably on a bracelet worn by Cady (Lindsay Lohan), she asks her what "fetch" means. Gretchen explains, "Oh, it's like slang, from England." She utters the phrase a final time in the context of a boy's compliment, which leads to an outburst from her friend Regina (Rachel McAdams): "Gretchen, stop trying to make 'fetch' happen! It's not going to happen!"

So what is it that makes certain new words "happen"? And why do many others, like Gretchen's, fail to do so? Slang terms are a bit like certain comets that appear suddenly in the linguistic firmament, glow brightly for a while, and then slowly fade away. In the early twentieth century, for example, the phrase "23 skidoo" made its appearance in the U.S. vernacular. Its origins are unclear, but it may have been the combination of two earlier, even more obscure expressions. Although it was never common, its popularity, in print at least, peaked around 1950 and then declined sharply until the mid-1960s. In the twenty-first century, when it is invoked at all, it often seems to be employed sarcastically—like "groovy"—to mock members of earlier generations who might have used the expression sincerely.

Ephemeral terms can create misunderstanding because there may not be widespread agreement about what they mean. Even the unabridged *Oxford English Dictionary* seems to be of two minds about "23 skidoo." One entry defines it as an "exclamation of disrespect" toward an individual, and three early twentieth-century quotations are provided in support of this interpretation. However, a second entry suggests that the phrase morphed into an imperative, meaning roughly the same thing as "scram." As evidence for this definition, four citations are given from the 1920s, 1950s, and 1970s. The Merriam-Webster dictionary seems to favor the "scram" interpretation and suggests that "skidoo" is a variant of "skedaddle."

It seems likely, therefore, that most people would be confused if someone said "23 skidoo" to them. Is it an insult or perhaps an order to leave? Is it intended ironically? As a meaningful expression, "23 skidoo" appears to have had its day and faded into historical obscurity. And this is undoubtedly the fate of a host of similar ephemeral expressions.

Other terms, however, enjoy a different career trajectory. They enter a language to take on specific roles. There are, after all, new things coming into existence all the time, and speakers need some way to refer to them. This can be accomplished in several ways. One straightforward solution involves borrowing words wholesale from other languages (as in "armoire" or "cologne"). The speakers of some languages engage in this behavior reluctantly, apparently fearing that exotic terms from distant parts may sully the purity of their native tongue. English, on the other hand, has always been fairly promiscuous in this regard. As James Nicoll vividly put it, "English has pursued other languages down alleyways to beat them unconscious and rifle their pockets for new vocabulary."[32]

A second possibility is that the speakers of a language cobble something together out of spare parts that are already lying around (as in "astronaut" or "polyvalent"). And sometimes, earlier terms take on completely new meanings, such as the names applied to six types ("flavors") of quarks: up, down, strange, charm, bottom, and top.

In other cases, a new term or phrase enters a language that already has a surfeit of expressions that mean the same thing. A good example would be the many ways of referring to being intoxicated. A list of dozens of such terms compiled in 1901 suggests that they tend to be ephemeral ("all mops and brooms," "more sail than ballast") or change their meaning throughout time ("feeling his oats," "has a bee in his bonnet").[33]

It has been difficult for researchers to study such neologisms because they tend to crop up suddenly in the vernacular as spoken slang. Terms like these appear in a written form—the raw material that lexicographers have traditionally trafficked in—only after they have already enjoyed some degree of success.

The rise of online social networks, however, has changed all of this. Social media provide scholars with a powerful new lens through which they can observe ephemeral terms as they are born and then either flourish or fade away. Social media posts are archived and date stamped. The number of views and "likes" that they generate are recorded. All of this allows usage and popularity to be tracked and quantified with an unprecedented degree of precision.

Consider, for example, the phrase "on fleek." On June 21, 2014, 16-year-old Kayla Lewis (aka Peaches Monroee) posted a six-second video on Vine in which she used those words to describe her perfectly groomed eyebrows. Both the video and the phrase went viral and spread rapidly to a number of other social media platforms.

By September, the phrase had begun to appear in the Twitter feeds of large corporations, such as Taco Bell, Denny's ("hashbrowns on fleek"), and IHOP ("pancakes on fleek"). A November Instagram post by Kim Kardashian employed the hashtag #EyebrowsOnFleek. By late 2014, it was appearing in the lyrics of artists like Nicki Minaj and Chris Brown. And in February 2015, it appeared in the *New York Times* in the title of a language quiz.[34] "On fleek" had clearly arrived.

The American Dialect Society nominated "on fleek" for its 2015 Word of the Year in the "most likely to succeed" category. The society defined the term as meaning "excellent, impeccable, and on point," which shows how its use had broadened to encompass more than outstanding eyebrows or pancakes. In a January 2016 vote by the society's membership, however, "on fleek" lost out to CRISPR, the name of a new gene-editing technology.

The votes cast by these scholars appear to have been prescient. Google Trends, which tracks the popularity of search queries, recorded "on fleek" as peaking in popularity in January 2015. It stayed strong through May of that year and then began a steady decline. On the other hand, searches for CRISPR surpassed those of "on fleek" during August 2016 and have continued to rise. By June 2019, searches for "on fleek" had fallen by 90 percent.

The phrase may stick around for a while—in 2019, it was the title of a song by the French artist Lartiste—or it may join the countless other ephemeral terms that came before it. And as researchers watch these phenomena play out in real time, they will be able to develop better theories about where new words come from and how their meanings mutate and evolve.

THE TROUBLE WITH TEXTING

As we have seen, some of the misunderstandings that arise from the use of electronic communication can be explained by a lack of shared common ground. For example, we typically send and receive email from a variety of people, including many who we may not know well. But a lack of common ground is a less likely explanation for miscommunication via text messages since they are typically exchanged by friends, family members, and romantic partners.

Shortly after texting became common, a study reported the results of focus groups held with college students. They were asked about misadventures they experienced when texting and why they thought miscommunication had occurred. Many of the respondents mentioned the brevity of texts as a primary reason for such problems.[35]

Text messages are typically brief because tapping on a smartphone with one's thumbs doesn't lend itself to long missives, and there are technological issues at play as well. SMS messages longer than 160 characters are broken up and sent as separate texts, and before unlimited text messaging became common, this could be expensive. For these reasons, many texts lack enough context to be interpreted successfully, as in this example:

Bob: It's over
Sarah: You're breaking up with me?
Bob: No the movie is over LOL
Sarah: LOL oops[36]

A more recent study suggests that even romantic partners struggle with this form of communication. People typically interweave their texting with many other activities throughout the day and sometimes respond with brief messages when they are busy. Short texts, however, are often perceived negatively by their recipients, who may mistakenly conclude that their partner is angry or doesn't care enough about them.[37]

In general, texting between romantic partners functions best when both parties are equally likely to initiate an exchange and send similar numbers of texts. Something as simple as texting to say hello is predictive of greater relationship satisfaction.[38]

The acronyms, initialisms, abbreviations, and other shortenings that are common in texting can be another source for confusion, especially between people belonging to different internet generations, as described in chapter 1. Some of these, such as ASAP ("as soon as possible") and FYI ("for your information") have made the leap onto smartphone screens from earlier forms of text-based communication.

Others, such as FWIW ("for what it's worth") and ROTFL ("rolling on the floor laughing") are of more recent vintage but widely used and commonly understood. Others, however, may be a bit more esoteric: examples might include BSAAW ("big smile and a wink") or ADIH ("another day in hell"). And some may have more than one interpretation, as in the following example:

Daughter: I got an A in Chem!

Mother: WTF, well done!

Daughter: Mom, what do you think WTF means?

Mother: Well That's Fantastic.[39]

Perhaps the most problematic, however, is LOL. It originally stood for "laughing out loud" (although British Prime Minister David Cameron apparently thought that it meant "lots of love" until he was corrected by Rebekah Brooks, the recipient of many texts from him).[40] Michelle McSweeney has argued that it is one of the older texting conventions and has evolved to serve a variety of functions.

At the end of a clause, it may simply be an intensifier, functioning like an exclamation point. It can also serve as a signal for sarcasm or flirtation. Its most common function, however, may be to soften a message that might otherwise seem too assertive, as in "Hurry up I'm starving LOL."[41] Given the wide variety of purposes that these three letters can serve, it's no wonder that LOL can cause confusion.

Finally, the technology itself should take some of the blame for misunderstandings. As mobile phones became more sophisticated, spell checking became a standard feature of text messaging software. On the one hand, autocorrect has undoubtedly saved most of us from making embarrassing mistakes when texting. On the other hand, these "corrections" can also create errors when it changes what we type into something very different than we intended. This phenomenon goes by several names, including "autofails" and the Cupertino effect.

Autocorrect errors are a staple on social media sites and have been published as book-length compilations.[42] They are popular because the substitutions are often bizarre, risqué, or both. Jerry Davich, a columnist for the *Chicago Tribune*, once wrote about an embarrassing message he sent to a female police officer. She had texted to ask about his arrival time and how he wanted to shadow her for the piece he was writing. He replied via audio text, stating, "I'll be there soon, and I'd like to ride with you." However, the message she received didn't contain the word "with," which changed the meaning of his reply considerably. Fortunately, the officer was not offended, and they were able to laugh about it when he arrived at his destination.[43]

Not everyone, however, is as forgiving. In northern England in 2011, a man named Neil Brook sent his neighbor a text that included the word "mutter." His phone, however, autocorrected "mutter" to "nutter," which is British slang for a crazy person. The recipient, one Josef Witkowski, was enraged by the insinuation that he was mentally ill and went to Brook's apartment, armed with a fake gun and a knife. Brook attacked him with two knives of his own, and after a struggle, he fatally stabbed Witkowski through the heart. Brook was tried for Witkowski's death and found guilty of manslaughter.[44]

Fortunately, most misunderstandings caused by emails or texts don't rise to the level of homicide. And perhaps part of the credit for that should be chalked up to the ingenuity of the online community as a whole. Faced with the communicative challenges of such impoverished media, people have created new cues, such as emoji, to disambiguate their messages. However, as we will see in the next chapter, they have also created new ways of being misunderstood.

6

MISINTERPRETING NONVERBAL LANGUAGE

Human communication is carried by much more than the words that we speak or write. We also rely on a host of nonverbal cues, such as facial expressions, gaze, posture, hand gestures, and even silence. Unfortunately, these nonverbal cues are as likely to being misinterpreted and misunderstood as the things that we say, if not more so. And the vagueness and ambiguity inherent in such displays has migrated to the online world as well. In that realm, we find emoji, originally intended to clarify texts and tweets, often doing the opposite of that.

(MIS)READING THE FACE

> *The countenance is an image of the mind, and the eyes are its interpreters.*
>
> —Cicero, *On Oratory and Orators*, LIX (55 BCE)

In agreement with Cicero, most people subscribe to the notion that emotions can be read in the face and that the eyes function as windows to the soul. The idiom "It's written all over your face" is commonly used and turns up in the lyrics of Ronnie Milsap, Jimmie Raye, and the Rude Boys, to name but a few of the artists who have employed it. But how true is this? Can the contractions of muscles in the face be read like the pages of a book? One obvious problem with this belief is that the expression of positive and negative feelings can have certain behavioral similarities: we shed tears in times of both great joy and great sadness. Perhaps it's not as easy to read someone else as it might seem.

It's also worth noting that our naive belief in reading faces can have real-life implications. Jurors in a mock trial, for example, required less evidence and were more confident in reaching a guilty verdict when shown a picture of an untrustworthy- versus a trustworthy-looking defendant.[1] It is clear that people are clearly influenced by facial cues. But even if such signs can be perceived on one's face, are people equally adept at reading them?

This question has received some attention by researchers, and it's been shown that many people have difficulties in determining someone's affective state based on their facial expression. Individuals with borderline personality disorder, which is characterized by impulsivity and problematic interpersonal relationships, seem to struggle with reading faces. They also show a negativity bias: a tendency to interpret an ambiguous facial expression in a negative way.[2]

This same negativity bias has been observed in those who suffer from social anxiety disorder, particularly after they have experienced social exclusion.[3] Emotion recognition deficits have also been discussed in the context of schizophrenia.[4] And compared to their younger counterparts, older adults are less accurate at identifying facial cues signaling fear or sadness.[5]

We might expect, therefore, that people on the autism spectrum would also struggle with facial emotion recognition since they often experience difficulty in interpreting social cues. However, the evidence for such difficulties is mixed: some studies have found no deficits in emotion recognition for people with autism, while other researchers have found severe impairment.[6]

Some facial displays are harder to read than others. Neutral expressions, in particular, can be problematic. For many people, their relaxed face is perceived as truly impassive by others. However, the human face at rest is highly variable in appearance, and for some, an expressionless visage seems to convey something else, like irritation. This phenomenon was given the unfortunate name "bitchy resting face" by the comedy group Broken People in an April 2013 video. Intended as a parody of a public service announcement, it quickly went viral on YouTube. By 2015, articles about the disorder were appearing in the *New York Times*, illustrated with pictures of women who seemed to suffer from the affliction, such as Kristen Stewart, January Jones, and Victoria Beckham.[7]

It should be noted that, despite the sexist name for this condition, resting bitch face can be a problem for males as well as females. In her book on impression management, Heidi Grant Halvorson relates the story of a team leader who would consciously put on his "active listening" face during meetings. After a few weeks of this, one of the team members asked if he was angry. This is how he discovered, much to his surprise, that his neutral listening face was perceived by others as irate.[8]

Some people are apparently willing to go to great lengths not to appear continually annoyed or hostile, such as by undergoing cosmetic procedures. Botox is a neurotoxic protein that, when injected, can make people appear younger by reducing the appearance of frown lines between the eyebrows. It was approved for this purpose by the U.S. Food and Drug Administration in 2002. However, some people who subject themselves to this treatment may be doing it to make their resting face look less, well, bitchy.[9] And it could be argued that possessing an eerily impassive face at all times is not much of an improvement.

As it happens, there are a number of medical conditions that can create abnormal facial expressions. One of these is Bell's palsy, a muscular weakness that can cause the features of one side of the face to droop. Another is Meig's syndrome, in which people involuntarily blink and thrust their chins. And among many other behavioral changes, individuals with Parkinson's disease typically develop blank, mask-like facial expressions.

So can we truly tell what another person is feeling by how they look? A growing body of research by psychologists suggests there is more variability and dependence on context in interpreting emotions than is commonly realized. The psychologist Lisa Feldman Barrett has become well known for arguing that behavioral displays associated with even basic emotions, like anger or happiness, can vary considerably from culture to culture.[10] If the face does provide a window into the soul, then the view it provides is not an especially clear one.

DO THE EYES HAVE IT?

When Irish eyes are smiling, sure, 'tis like the morn in spring.

—Chauncey Olcott and George Graff Jr. (1912)

During her run as the host of *America's Next Top Model*, Tara Banks introduced a number of terms into the American vernacular. These include portmanteau words like "sinnocent" (a combination of "sexy" and "innocent") and "flawsome" (embracing one's flaws). Most of these neologisms failed to catch on, but one, to "smize," or smile with the eyes, became reasonably popular, although as of this writing, it hasn't made it into the Oxford or Merriam-Webster dictionaries. Banks appears to have started using the term in 2009 during the thirteenth season of her show.[11]

We typically think of smiling as an act performed by the mouth. How is it possible, then, to communicate a smile via the eyes? Until recently, the question was mostly academic. After all, most of us aren't top models, and our conversational partners can typically see our mouths as well as our eyes.

But the COVID-19 pandemic and the attendant wearing of facial coverings created one of the greatest communication experiments in history. Face masks cover the nose, mouth, and chin, leaving only the eyes and eyebrows to communicate affective responses. Can a smile still be communicated when half of one's face is covered, or is this yet another vector for miscommunication? To put it simply, is smizing really a thing?

To answer this question, we need to consider the physiology of the face and the way that our muscles contract during displays of emotion. Smiling doesn't seem to be learned; instead, it is an innate reflex to a variety of stimuli, such as being stroked or in response to particular odors. And social smiles—grins in response to other people—start when infants are four weeks of age, on average.[12]

Smiling involves the contraction of a number of muscles in the face. And there are two different kinds of smiles. In a genuine smile caused by happiness, the zygomaticus major muscles raise the corners of the mouth, while the orbicularis oculi muscles raise the cheeks, which causes a crinkling of the skin around the eyes. (This creates the characteristic creases known as crow's feet or laugh lines.) Genuine smiles are often referred to as Duchenne smiles, after the nineteenth-century French neurologist Guillaume Duchenne.

But we can also smile when we're not particularly happy, as when we're asked to do so for a photograph. These contrived smiles, however, involve the zygomatic muscles only, as the muscles that raise the cheeks are not under voluntary control for most people. And most of the early research on smiles suggested that Duchenne smiles are perceived as more genuine than posed smiles.[13]

More recent research, however, suggests that a majority of people can produce posed smiles that are perceived as genuine as well.[14] This is because a slight squint (or "squinch," as photographer Peter Hurley has dubbed it) can create the same lines around the eyes as the movement of the muscles that raise the cheeks.[15]

Assuming that people are able to smile with their eyes, how good are others at detecting such facial expressions if all they can see are someone's eyes? In the 1990s, British psychologist Simon Baron-Cohen devised a test to assess this. He cropped pictures of people's faces so that only the eyes and eyebrows could be seen. He then asked research participants to choose which of two words was the better descriptor of what the person in the picture was feeling or thinking. For example, the words might be "serious" or "playful."

The test has been widely employed as a way of determining whether a person has a fully formed theory of mind. This refers to one's ability to attribute mental states, such as emotions, to another person. Individuals diagnosed with autism tend to have theory-of-mind deficits, and they also perform more poorly than non-autistic individuals at "reading the mind in the eyes."[16]

There are, however, differences even among neurotypical individuals. Studies have shown that females outperform males on this task.[17] Age seems to play a role as well: People in their forties and fifties are the most accurate at this task and perform much better than teenagers or those in their sixties.[18] It should be noted, however, that the test has been criticized as having a number of biases. Highly educated white participants tend to outperform those with less education and those who belong to other races.[19]

The eyes are important in communicating emotion, but even the eyebrows get in on the act. They can also be expressive and will change their shape to the accompaniment of a smile. Although they may spontaneously rise, as when people are surprised, they are also under voluntary control. Flexing the eyebrows up and down, à la Groucho Marx, can communicate mischievous intent. Body posture and movement play a role in communicating affect as well.[20]

Although it is possible to perceive someone's friendly demeanor when the lower part of their face is obscured, doing so can sometimes be a challenge. A mask counts as a strike against effective communication and can make it harder to decipher more complex displays, like disgust or sadness.[21] During the pandemic, many people found themselves exaggerating their expressions to make them more apparent. We may not be able to read our conversational partner's minds, but interpreting what they are feeling is easier when both the eyes and the mouth can be seen.

TOKEN GESTURES

As we have seen, divining the meaning of words and phrases can create a multitude of problems. And if anything, the interpretation of symbols is even more fraught. After all, one can always consult a dictionary to get some sense of a word's definition and its usage. Compilations of signs and symbols do exist, but they may not be particularly helpful.

Conducting a web search for "pitchfork," for example, will return millions of matches, and this may not provide much assistance in determining the meaning of a pitchfork that someone has spray painted on a wall near your favorite restaurant. Is it a meaningless graffito, or does it represent something more significant? And what should we think when someone gesticulates in a way that we don't understand? Let's begin by considering the case of gestures.

Many travelers have had the misfortune of learning that hand gestures can mean distinctly different things in different countries. Consider, for example, the upturned thumb. It is a gesture of approbation in some countries but an insulting one in much of Latin America, West Africa, and the Middle East.

Even within the same culture, however, a gesture can mean different things. Among scuba divers, a thumbs-up means "up" or "end the dive." A diver might use it to signal to others, for example, that he is uncomfortable about continuing with a dive. And the thumbs-down gesture signals that a diver is ready to continue and go deeper. In both cases, the thumb gestures mean the opposite of their terrestrial counterparts.

In other cases, the meaning of a gesture depends on the context in which it is employed. Extending one's thumb and little finger outward, for example, is the shaka symbol ("hang loose," or take it easy) and has frequently been employed by President Obama, who was born on Oahu and lived there for several years. However, when the same gesture is made next to the face, with the thumb near the ear and the little finger next to the mouth, an entirely different meaning is intended. This has become the universal sign for "call me," with the shape of the hand mimicking the receiver of a traditional telephone.

Other gestures have morphed throughout time. The raising and splaying of the index and middle fingers to form a V was the universal "victory" symbol used by the Allies during World War II and popularized as part of the BBC's "V for Victory" campaign. Winston Churchill was an enthusiastic adopter of the gesture, and it went on to become emblematic of victory in politics as well. It was used in this way by President Eisenhower in the 1950s and then by his vice president, Richard Nixon, in the 1960s and early 1970s. Politicians have typically made the gesture with both hands and upraised arms. Weirdly, Nixon also made this gesture after resigning in 1974 as he boarded a helicopter to fly into exile.

When the V sign is made with the palm facing inward instead of outward, however, it is intended as an insult—at least in the United Kingdom and many Commonwealth countries. Often delivered with an upward motion of the hand, it is the equivalent of saying "up yours." Interestingly, in the many photographs of Churchill making the V for victory symbol, he is not infrequently seen making the gesture with his palm facing inward. Some have speculated that, given his aristocratic background, Churchill was simply unaware of the derogatory meaning. Others have suggested that he knew exactly what he was doing and that he was communicating two messages at the same time: a V for victory to his countryman and allies and "up yours" to the Axis powers.[22]

During the 1960s, the V sign was appropriated by Americans opposed to the war in Vietnam and became known as the peace symbol. The gesture was a potent symbol of the counterculture and was widely employed at antiwar demonstrations. It has been claimed that it originated in 1966, when Emmett Grogan, who belonged to a counterculture group in San Francisco, was released from jail. He made the insulting British gesture when he noticed that a news photographer was taking his picture. When this photo appeared in the newspaper on the following day, his friends adopted it, mistakenly believing that Grogan had intended the gesture as a victory sign.[23] Given its militaristic origins during World War II, its adoption by protesters may seem a bit ironic, but both the soldiers of the early 1940s and the antiwar demonstrators a generation later were advocating for the same thing: an end to war.

The V sign continues to evolve. It is a gesture often made in East Asia by people posing for selfies or group photographs. There are various theories about its origin: some link it to the V for victory sign, while others suggest that its popularity is associated with its use by celebrities. Among women in Japan, the gesture has become associated with *kawaii* ("cuteness").[24]

The raising of the index and little finger is a gesture frequently seen at concerts. Its link to heavy metal was popularized in the late 1970s by Ronnie James Dio, best known as a vocalist for Black Sabbath. The gesture, often referred to as the sign of the horns, had been used by Dio's Italian grandmother to ward off the evil eye and became known as "maloik." Dio is thought to have adopted the gesture to differentiate himself from Ozzy Osbourne, the band member that he replaced and who frequently made the double peace sign during his performances.

The same gesture has been used since 1955 by fans of the University of Texas Longhorns. The two outstretched fingers approximate the prominent horns of their namesake cattle, and the phrase "Hook 'em Horns" is associated with the gesture. In other parts of the world, however, the sign is associated with cuckoldry (as in Italy) or a salute to Satan (as in Scandinavia). Many European news outlets expressed astonishment when George W. Bush, as well as his wife and daughters, made the gesture during the parade for his second inauguration in 2005. Bush was, however, saluting the Longhorn marching band and the alma mater of his wife Laura and daughter Jenna.[25,26]

A gesture similar to the sign of the horns is associated with Gene Simmons of Kiss, although in his version, the thumb joins the extended index and little fingers. He even tried to trademark it. This is the same gesture as the ILY, or "I love you," sign in American Sign Language: it is the combination of the letters I, L, and Y in the manual alphabet. It was popularized by figures as diverse as game show host Richard Dawson, President Jimmy Carter, and professional wrestler Jimmy Snuka. And fans of the University of Louisiana Ragin' Cajuns utilize the same sign as an initialism for their school, UL. There is, however, a darker side to the use of gestures and symbols, and that is the subject of the next section.

SIGN, SIGN, EVERYWHERE A SIGN

As most people are aware, street gangs use various signs to signal affiliation and to mark the territory they control. And since gangs are associated with crime and violence, many parents worry that their children may become involved in gang activity. Schools have become a flashpoint for such concerns since they are responsible for ensuring the safety of their charges for several hours each day.

At a high school near Sacramento, for example, a group of seniors who were to graduate in 2014 ordered sweatshirts with "XIV" emblazoned on them.[27] This seems innocent enough, but the number 14 is associated with Norteños, a group of gangs active in Northern California (the letter N is the fourteenth letter of the alphabet). As a result, the students were advised not to wear the shirts.

More significantly, a black 15-year-old student was expelled from his high school in Mississippi for being photographed holding up three fingers. This corresponded to the number on his football jersey, which he was wearing in the photo. Unfortunately for him, however, holding up the middle and index fingers (forming a V) and thumb (forming an L) is a symbol of the Vice Lords gang.[28] The student's expulsion was triggered by the school's zero tolerance policy. And parents pressured a Virginia high school principal to resign after a photo surfaced that showed him making an ambiguous gesture beside five students who were also making hand gestures. The principal claimed that he had been motioning for the students to stop making their own gestures when the picture was taken.[29]

But it's not only students and school officials who have run into controversy with regard to ambiguous hand signs. In 2014, Minneapolis TV station KSTP ran a picture of mayor Betsy Hodges posing with Navell Gordon, a black community volunteer, during a get-out-the-vote drive. Both individuals are awkwardly pointing at each other. This was interpreted by the station's reporters and by a retired police officer as a "known gang sign." #Pointergate, as it became known, was widely derided by national publications such as *Vanity Fair*[30] since it appeared to be a particularly egregious example of racial profiling.[31]

And what could be more innocent than the heart sign, in which the thumbs and index fingers are curled together into a symbol of love? In 2007, the Virginia Tourism Agency developed a campaign called "live passionately" as part of their well-known "Virginia Is for Lovers" advertising. The ads included actors making the heart symbol in a variety of contexts, including—improbably enough—while stomping grapes. However, it was pointed out to state officials that the heart sign was also a gesture employed by the Gangster Disciples, and as a result, the ad campaign was scrapped.[32]

Gangster Disciples' symbolism also proved problematic for Urban Outfitters. The company discontinued its Vanguard Pitchfork Tee in the summer of 2013 when a resemblance to the pitchforks that appear in the gang's symbol was noted.[33]

Even the "okay" hand gesture—the symbol for approval and approbation—has become controversial. In early 2017, members of the internet forum 4chan flooded the web with claims that the hand sign was a symbol representing white power: the three raised fingers represent the letter W, and the circle made by the index finger and the thumb stand in for the top of the letter P. Even though this was a transparent attempt at trolling, a couple of Chicago high schools decided to edit and republish their 2019 yearbooks when it was noticed that they contained pictures of students making the gesture.[34]

Concerns about this gesture and its meaning made headlines at the end of 2019. Before the December 14 Army–Navy football game, the television audience saw cadets and midshipmen repeatedly making a hand gesture behind Rece Davis, the ESPN commentator. Although it appeared to be the familiar "okay" sign, there was speculation that it was being employed as a symbol of white supremacy. The incident took on a political dimension because Donald Trump had met with the players beforehand and was present during the game. An investigation by the academies concluded that the gesture was part of a prank called the circle game. The goal is to trick someone into looking at the hand gesture; if they do so, the loser can be punched by the winner.[35]

The issue came to the fore again in April 2021, when Kelly Donohue, a *Jeopardy!* contestant, held up fingers to mark each of his victories. After his third, some fans of the show thought that, by holding up three fingers, he was flashing a flattened version of the white power gesture. Even though Snopes and the Anti-Defamation League concluded that it had been an innocent gesture, some of the show's devotees felt otherwise.[36]

The hypervigilance surrounding signs and their meaning has fallen particularly heavily on the deaf community. Most people are not familiar with sign language and therefore may misinterpret these hand gestures as something else, such as the flashing of gang signs. In 1995, an altercation at a Minneapolis bus stop led to a deaf man having a beer bottle broken over his head and his eyes gouged with the bottle's shards. His offense? The assailants had interpreted his sign language as gang signs.[37]

And in 2011, two men who were conversing in sign language at a bar in Florida were attacked by a knife-wielding assailant who mistakenly thought they had been exchanging gang signs.[38] An eerily similar incident occurred 20 months later in North Carolina. Two men were engaged in a sign language conversation while they walked down the street. They were attacked by a woman with a kitchen knife, and one of the men was stabbed several times.[39]

It seems fair to say that concern about gang activity has created something of a moral panic. And as we have seen, this unease seems to be spilling over to anxiety about members of other groups, such as white supremacists. As a result, people seem inclined to perceive ambiguous or unfamiliar communicative gestures as potentially threatening and respond accordingly.

WORTH A THOUSAND WORDS?

Emoji are the communicative descendants of emoticons, those winking and smiling "faces on their sides" that were used to help disambiguate email messages. Emoticons have now been largely supplanted by emoji: thousands of small icons that provide representations of facial expressions as well as many objects, animals, plants, and whimsical combinations, such as smiling cat faces with hearts for eyes.

One of the most popular emoji is "face with tears of joy" (U+1F602); it was named 2015's "word" of the year by the Oxford dictionary.[40] Its usefulness may be due to the fact that such emoji can function as relationship maintenance tools, a point made by psychologist Monica Riordan. She suggests that, in many cases, they can be used as signals of affection.[41]

With regard to miscommunication, however, a primary issue regarding their use is a lack of consensus regarding what they truly mean. An organization called the Unicode Consortium is responsible for standardizing the character encoding schemes that are used by smartphones, web browsers, and other software. Thanks to the consortium's work, U+1F63B is the hexadecimal code that will always depict a smiling cat face with hearts for eyes regardless of the user's platform. This organization, however, plays no role in dictating when or how such an emoji should be employed.

Another problematic issue is that software developers have some leeway in determining how a particular emoji will be rendered on a given device.[42] Our "cat with hearts for eyes," for example, will appear somewhat differently on an Apple iPhone running iOS compared to Google's Android or in an app like Facebook or WhatsApp.

Across platforms and programs, the cat's whiskers, degree of smiling, and size of the "hearts for eyes" will vary, sometimes significantly. (The range of possibilities for any character code can be found at the online resource emojipedia.org.) These variations may be the result of intellectual property concerns, but they create yet another vector for misunderstanding.[43]

Although the meanings of words can be found in dictionaries, no one decrees what a particular emoji stands for. The good news is that an emoji can mean almost anything. And the bad news? An emoji can mean almost anything. Emojipedia.org does sometimes provide usage notes. For "face with tears of joy," for example, it notes that it is "widely used to show something is funny or pleasing." Puzzled recipients of cats with hearts for eyes, however, are left to their own interpretations.

The ambiguity created by the use of certain emoji has been a topic of discussion for some time. A good example would be "folded hands" (U+1F64F). This can be understood as meaning "thank you" or "please" in Japan[44] or as a greeting in India. In the West, it has often been interpreted as praying hands. However, some believe that it depicts the "high five" gesture, with the hands of two different people coming together in celebration.[45]

And some otherwise unremarkable emoji have become infamous because they have been appropriated to symbolize various parts of the human anatomy. Although there are emoji for various bodily appendages and organs, there are none for the buttocks or the phallus. As a result, people have pressed other emoji into service to represent them. The peach and the eggplant emoji are well known stand-ins that are often used, for example, in sexting.

Starting in September 2019, Facebook banned the use of emoji for sexual purposes. Specifically, Facebook's Community Standards, Part III, Section 15, prohibits "content [that] facilitates, encourages or coordinates sexual encounters between adults." This prohibition includes "contextually specific and commonly sexual emojis or emoji strings."[46]

The peach and the eggplant aren't mentioned by name, although most commentators have interpreted Facebook's edict as including them.[47] But given that the peach emoji also became popular in 2019 and 2021 as shorthand for im*peach*ment, it becomes harder to see how such a ban could be reasonably enforced. The situation became even more muddled in January 2021, when Georgia Democrats celebrated their state's senate victories by tweeting and texting the peach symbol. If Freud had been born in the 1950s instead of the 1850s, he might well have observed that, sometimes, an eggplant is just an eggplant.

Research studies have also documented a lack of consensus about the meanings of emoji. A study conducted in 2016 asked participants to categorize various emoji as having positive, negative, or neutral meanings. The raters disagreed with one another a quarter of the time. Of the

15 different emoji that the researchers asked about, the "smirking face" (U+1F60F) led to the greatest number of divergent responses, with participants indicating that it denoted disappointment or being dismayed, unimpressed, depressed, or indifferent. Many of the participants also pointed out that emoji render differently on different platforms and devices, as was mentioned earlier.[48]

One could object that the researchers were asking about emoji in the absence of context. However, a later study queried participants about the meaning of emoji when they appeared alone and also with some verbiage to provide a frame of reference. The additional information didn't seem to help. When asked why, several participants said they couldn't rule out a sarcastic interpretation.[49]

Attorneys and judges have struggled with how emoji should be presented as evidence in the courtroom. During a high-profile 2014 trial, for example, a lawyer requested that his client's chats, forum posts, and emails be shown to the jury instead of read out loud. His concern was that the meaning of emoji "cannot be reliably or adequately conveyed orally."[50]

Finally, there are shifting standards with regard to the appropriateness of using emoji. A 2015 article in the *New York Times* suggested that men, in particular older men, shy away from using them.[51] Since that time, however, they have become even more ubiquitous and are another example of how the internet has made informal language more acceptable.[52]

THE PLAY'S THE THING

In athletic competition, communication between the members of a team is vital. In addition, the coach of a team needs to be able to call offensive plays. Let's consider how these communicative goals are accomplished in two different games: baseball and football.

In baseball, the catcher signals the pitcher to provide advice about which pitch to throw based on factors like the attributes of the current batter, the count, and how many runners are on base. The catcher accomplishes this by pointing downward, between his legs, with one or more fingers of the throwing hand, which is held close against the body.

Since the catcher is squatting, his legs block any view of the sign by members of the opposing team.

In baseball, however, it's permissible to steal signs, and there is a long and venerable history of such activity. This becomes possible, for example, when there is a runner on second base since he is standing behind the pitcher and can see the catcher's signs. The base runner can attempt to communicate this information to the batter.[53]

Although such sign stealing is permitted, players and coaches are not allowed to use any mechanical device—such as binoculars or cameras—to steal signs.[54] In 2017, for example, the Boston Red Sox were caught cheating when they were provided with catchers' signs by video replay technicians and then passed them along via Apple watches.[55] And that year's championship winners, the Houston Astros, used both high-tech (video cameras) and decidedly low-tech (banging on trash cans) methods to relay the opposing catchers' signals to their own batters.[56]

The third-base coach also provides guidance via signs. He might, for example, signal for the batter to bunt or encourage a runner to steal a base. Unlike the catcher, however, a third-base coach's signs are visible to all, and therefore he will employ decoy signs—ones that are meaningless to the batter and runners—before making a gesture that signals a real instruction. He might, for example, slide a hand down a thigh, scratch his face, and brush at a shoulder before producing a genuine sign, such as touching his cap to indicate a bunt. Therefore, these signs often appear more complicated than they truly are.[57]

Since the number of plays requiring signs is relatively low, this system works well for baseball. However, it's possible that a player who is traded might reveal the signs that were used by his previous team, and a third-base coach may change his signs after a trade occurs.[58]

Even with a simple system, mistakes still occur. Inattention by a batter or a base runner can cause a signal to be missed. In addition, a wrong signal might be made. In June 1991, the Mets lost a game to the Braves because Mike Cubbage, the third-base coach, signaled a risky hit-and-run play. In a hit-and-run, a first-base runner will start moving as soon as a pitch is thrown in an attempt to steal second. This causes the infielders to move in turn, and then the batter tries to hit the ball into one of the resulting infield gaps.

In this case, the Cubs' manager, Bud Harrelson, had signaled to Cubbage that Vince Coleman, who was on first base, should simply be encouraged to steal second. Cubbage, however, accidentally made the hit-and-run sign instead. The batter, Dave Magadan, failed to make contact with the ball, and Coleman was thrown out as he attempted to steal second.[59]

Communication is more complicated in football because the number of offensive plays is much larger, and the plan must be disseminated to all the players on the field. For a particular game, there might be 80, 90, or even 100 pass plays and as many as 20 running plays. To deal with this formidable memory load, quarterbacks have taken to wearing wristbands that have dozens of plays inscribed on them.[60] The quarterback might choose to alter the play—that is, "call an audible"—if he believes that the defense has guessed which play is about to be run and have arranged themselves accordingly.

In the NFL, quarterbacks and middle linebackers receive play calls from their coaches by means of radios in their helmets. This is not allowed in the college game, however. To play up-tempo, no-huddle offense, college teams have adopted play cards or boards for rapid communication between coaches and players.

Play cards were first employed by Oklahoma State in 2008. These large signs might have four or six sections, festooned with an apparently random assortment of pictures, cartoons, and other symbols, such as emoji and internet memes. These play boards can be used in conjunction with wristband card systems. As with any other public system, however, the meaning of the signs can be guessed at or stolen, and as a result, the systems are constantly changing. Decoy signs are also employed.[61]

Has the use of play cards cut down on the miscommunication of plays on the field? It's difficult to say since teams tightly control any information about the systems they employ. Visual systems like play boards would seem to be superior to audible signals since the roaring of the crowd can make it difficult for teammates to hear each other on the field—sometimes deliberately so.

In September 2019, University of Kansas receiver Andrew Parchment took responsibility for losing to Coastal Carolina because of a misheard signal that caused him to run the wrong route. As quarterback Carter

Stanley put it, "It was two words that are pronounced very similar, and they mean completely different things."[62]

On the other hand, changes made to play board systems require players to commit these changes to memory, and systems that continually change may be challenging to remember. Miscommunication can still occur—even when someone is waving large, colorful signs on the sidelines.

THE BOUNDS OF SILENCE

> *In human intercourse the tragedy begins, not when there is misunderstanding about words, but when silence is not understood.*
>
> —Henry David Thoreau, *A Week on the Concord and Merrimack Rivers* (1849)

> *I shall assume that your silence gives consent.*
>
> —Socrates, in Plato's *Cratylus* 453a

On August 29, 1952, the American composer John Cage premiered a new work to concertgoers in Woodstock, New York. As part of a recital, pianist David Tudor sat at his instrument for four and a half minutes. Some sort of performance appeared to be taking place: Tudor opened and closed the piano's lid and employed a stopwatch to observe the duration of each of the work's three "movements." But he didn't play a single note.

The reaction of the audience to Tudor's performance was variable. Some were merely puzzled, whereas others became uncomfortable. As more time went by, some of the patrons became angry and walked out of the hall.[63] Cage intended his work, *4'33"*, to be a meditation on the sounds both within and outside of the concert hall. But the audience's reaction seemed to be the result of violated expectations about what should occur during a concert as well as a general discomfort with the experience of a silence lasting several minutes.

Sixty years later, American college students did not fare much better than Cage's audience. When psychologist Timothy Wilson and his collaborators gave research participants the choice of spending 15 minutes alone with their thoughts versus giving themselves electric shocks, two-thirds of the male participants and one-quarter of the female subjects chose to self-administer the painful jolts.[64]

As has been amply demonstrated in previous sections, the things that people say can lead to misinterpretation and miscommunication. However, as Thoreau and Plato remind us, an absence of words and the silence that results can be even more problematic.

Before going further, it should be noted that there are cultures in which silence is, if not golden, then at least less traumatizing than electric current. Traditionally, Japan has been described as a country in which its inhabitants "tilt toward silence." Nonetheless, these silences can mean different things in different situations. They can, for example, signal that something is believed to be true, but they can also reflect embarrassment or even defiance.[65]

Periods of silence also vary in duration. At one extreme, there may be only a slight pause, as when someone gathers her thoughts or searches for the right way to say something. These are typically not problematic. And moments of silence might be taken by a group as an act of remembrance or commemoration. These institutionalized instances of "situational silence" are typically one or two minutes in length and are explicitly demarcated.[66]

At the other extreme, however, a silence may continue for an extended period of time, and its duration may not be known beforehand, as was the case for *4'33"*. And it is these situations that many experience as uncomfortable or anxiety inducing.

It appears that teachers are uneasy with even short periods of silence following their questions to a class. One classic study, for example, found that third-grade teachers of reading waited for only about a second for their students to respond to a query.[67] By moving on to an easier question or supplying the answer, however, teachers miss the opportunity to determine the source of their students' lack of response. This also occurs despite the fact that a longer wait time seems to be beneficial for students.[68]

Within the context of a conversation, the meaning of silence can be problematic. Listeners tend to view pauses at the end of a speaker's utterance as a signal that she is surrendering the conversational floor to someone else. To prevent that from happening, speakers will often insert a meaningless particle, such as "uh" or "um," into any silences that occur in their speech.[69] These are intended as signals to ward off interruptions and to buy the speaker some time to finish expressing her own thoughts. The hostile takeover of a conversation can occur at any juncture, but doing this while someone is in the act of speaking is typically perceived as particularly rude.

The meaning of silence from one's conversational partner can often be disambiguated by their nonverbal behavior. A pensive expression and an upward gaze, for example, can explain why someone has not responded: they need a moment to think. On the other hand, wide eyes and a mouth falling open provide evidence that someone has been stunned into silence, perhaps by a piece of unexpected news.

This helps to explain why it can be difficult to determine the meaning of a pause during a voice-only phone conversation. Considerate communicative partners will provide some explanation for their delay in responding to ensure that the other party doesn't feel neglected.[70]

Given the multifarious nature of silence, it shouldn't be surprising that it often leads to misunderstandings. In twenty-first-century America, most people would strongly object to Socrates' assertion that silence "gives" consent. However, as a character in Plato's dialogue, Socrates was only looking for a reason to continue with his line of reasoning. And although there are various forms of consent, most people probably think about it in the context of permission to initiate activity of a sexual nature.

Traditionally, some sort of nonverbal consent was the norm; that is, if one's partner didn't object, then they were implicitly agreeing to a sexual experience. This is embodied in the legal precept *Qui tacet consentire videtur* ("silence implies consent"). In the wake of movements like #MeToo and Time's Up, however, the default has clearly shifted to affirmative consent: the person initiating sexual contact is expected to request explicit, spoken permission before proceeding.[71]

7

COGNITIVE FACTORS

O ur minds construct meaning from language, but this complex process of interpretation sometimes goes awry. For example, the sentences that we try to comprehend can have more than one meaning. In addition, even a seemingly trivial change like the placement of a comma can radically alter how something is understood. And our memory can play tricks on us and create confusion when two concepts have conceptual similarities.

CHINESE WHISPERS AND RUSSIAN SCANDALS

"Somebody thinks of a sentence. He passes whatever he says on to someone else. That person says what he heard to someone else and so on and so on until it goes to the last person. Before the person who said the sentence (at first), the last person says what he heard. It has often completely changed."[1]

The game described in the preceding paragraph is popular throughout the world and is known by a variety of names. In the United States, it is usually referred to as Telephone, but in the United Kingdom and Commonwealth nations, it's best known as Chinese Whispers. Other names include Russian Scandal, Whisper Down the Lane, Secret Message, and the Messenger game.[2] The French play *téléphone arabe*.[3] In other languages, the name often translates to "broken telephone."

But why is Telephone played only by children as opposed to people of all ages, like charades? The answer may be that adults are already familiar with the unreliability of messages transmitted through the grapevine. For example, they may have been the victims of gossip or rumors that were spread by their peers in school. Younger children, on the other hand, can benefit from being explicitly taught about the distorting effects of second- and thirdhand information. They need to discover that garbled messages are yet another strike against effective communication.

It's probably safe to say that young children experience their fair share of miscommunication. For example, they often hear words they are unfamiliar with or grammatical constructions they have not yet mastered. At the same time, their vocabulary may be insufficient to clearly express their needs, wants, and desires. As a result, children don't need to master the *concept* of miscommunication: it's already a common part of their everyday experience.

The anthropologist Elinor Ochs has examined the different ways that caregivers deal with children's unintelligible utterances. For example, an adult might choose to ignore the child's utterance, simply acknowledge it, or guess at what the child intends. Ochs believes that these responses are not simply linguistic choices but instead reflect social and cultural factors that may vary from place to place. American mothers, for example, often guess at what their children may be trying to say, but such a response is far from universal.[4]

Research suggests that even young children are capable of recognizing communication failures and that they even have some ideas about how to address them. For example, one- and two-year-olds will repeat themselves when they are not understood and may try to articulate the words they are speaking more clearly. They are also adept at using gestures to reinforce or augment their meaning.[5]

Laboratory studies have been conducted in which toddlers, after requesting a particular object or action, were told "I don't know what you mean" by an adult. In these situations, the children proved capable of revising their requests. For example, one of the participants, when confronted with a lack of understanding after saying "open," changed the request to "want off." These episodes of repair suggest that even preschoolers have an understanding of how to alter their requests to make themselves understood.[6]

Preschoolers are even able to cope with episodes of miscommunication when talking among themselves. A study conducted at a kibbutz found that two- and three-year-olds are capable of producing clarification requests when their peers didn't understand them. The results showed, however, that children younger than three are less likely to produce such clarifications. This suggests that it may take some time to develop competence in creating conversational repairs.[7]

It may be that some children are simply better at conversational repair than others, and one study suggests a possible reason for this. Sara Bacso and Elizabeth Nilsen had four- to six-year-olds play a game in which they provided descriptions of pictures. When their descriptions were ambiguous, they received feedback that they had been misunderstood. The children's executive functioning was also assessed via tasks that measured their working memory capacity, cognitive flexibility, and inhibitory control. The children with greater cognitive flexibility proved to be the most adept at repairing their ambiguous statements.[8]

Young children also seem to understand that some informants are more trustworthy than others. Three-year-olds make such judgments in a relatively inflexible way: they are likely to mistrust anyone who has been proven wrong, even if it only happened once. Four-year-olds, however, seem to be more forgiving and will decide who they trust based on the accuracy of an individual throughgout time.[9]

There are, however, certain kinds of linguistic activities that children may not have much experience with, and one of these is obtaining information through long communication chains. Playing Telephone can teach them several important lessons about the hazards of such chains.

One lesson is that a message can be unintentionally distorted because it's hard to make out, as in trying to understand someone who is whispering into your ear. The medium of transmission can also degrade a message: think of trying to hear someone on a poor cell phone connection. And crucially, the game of Telephone demonstrates that the more parties a message must pass through, the greater the likelihood for misunderstanding or miscommunication.

Even if a child never plays Telephone, he or she will absorb the lessons of the game as they experience these sorts of communication breakdowns in real life. And once such lessons have been learned, the game ceases to be of much interest.

GOING DOWN THE GARDEN PATH

Time flies like an arrow; fruit flies like a banana.

—Usenet newsgroup net.jokes (1982)

When people hear or read sentences, they need to figure out who is doing what to whom. Different languages do this differently. In highly inflected languages like Latin, for example, roles like subject and object are marked by suffixes placed on nouns. In a language like English, the heavy lifting is typically accomplished via word order. In either case, the task of the comprehender is to make sense of the referents through a process called parsing.

The act of parsing is more difficult if sentences can be interpreted in more than one way. This creates what researchers refer to as syntactic ambiguity. Consider one of the well-known examples offered up by George Miller and his colleagues:[10]

They are cooking apples.

This sentence can be interpreted in two distinctly different ways. According to one reading, it means that some people are in the process of cooking apples as opposed to cooking something else. However, it can also be understood as referring to apples that are only good for cooking. Ambiguity is created because, from a syntactic perspective, "apples" can function as either the subject or the object of the sentence, with "cooking" functioning as either an adjective (modifying the noun "apples") or a verb (what is being done to the apples).

In spoken language, syntactic ambiguity can be avoided by applying stress to one of the words in the sentence. "They are cooking APPLES" would guide the listener to the first interpretation described above, whereas "They are COOKing apples" would be consistent with the second meaning.

Sentence-level ambiguity can manifest itself in other ways. Consider this:

Navy Ships Head to Puerto Rico

This may seem like a perfectly ordinary news headline, and it ran above an Associated Press story in April 2000. But a moment's thought will show that it, too, can be interpreted in more than one way. Both "ships" and "head" can function as nouns and as verbs, and in an alternate reading, one could imagine the navy sending a human head, perhaps via FedEx, to San Juan. In this case, however, the alternate interpretation is more than a little unlikely.

In other cases, the parsing process seems to fail completely on a first attempt. Consider this frequently discussed example:

The horse raced past the barn fell.[11]

Most people who read this sentence reach the word "fell" and become confused. They are forced to return to the beginning to try to reinterpret what they've read. It almost feels like being deceived. One idiom for such deception is a "false scent," and this is the term that H. W. Fowler used to refer to such constructions.[12] Another is "being led down the garden path," and researchers refer to such instances as garden path sentences.

Why does this confusion happen? It has been argued that our mental parsers prefer "early closure"; that is, the parser tries to identify a constituent noun phrase and close it off as early as possible in the interpretation process.[13] In most cases, this leads to a correct interpretation, and it also lowers the overhead of keeping long phrases active in working memory.

However, such a strategy won't work for comprehending garden path sentences. If the parser interprets "The horse" as a noun phrase, closes it off, and moves on to identify the next constituent, the process will fail. Because the sentence contains a reduced relative clause, the phrase "raced past the barn" goes with the horse mentioned at the beginning (as in "The horse that was raced past the barn fell").

We encounter sentences with reduced relative clauses more often than you might think because such function words are often omitted in newspaper headlines if space is at a premium. As a result, they are more likely to be grammatically impoverished and more difficult to understand than unambiguous sentences. Consistent with this notion, researchers have found that ambiguous headlines take longer to read than those with only one interpretation.[14]

Researchers have also studied ambiguous headlines for their humor value. As we saw earlier with the navy ships example, multiple meanings may be the result of syntactic ambiguity, but others are due to verbal or other forms of lexical ambiguity ("Kids Make Nutritious Snacks"; "Farmer Bill Dies in House").[15]

Ambiguous headlines are referred to as "crash blossoms." They were given that moniker by Dan Bloom. The odd name is an homage to a headline spotted in the newspaper *Japan Today* that read "Violinist Linked to JAL Crash Blossoms." Bloom found himself wondering what a crash blossom was until he realized that reference was being made to the violinist's successful career.[16]

Making sense of headlines found online can be even more problematic since they often appear without disambiguating context, such as a subhead or a photograph.[17] In the case of "Kids Make Nutritious Snacks," for example, a picture of a smiling child spreading hummus on bread would probably be enough to prevent visions of cannibalism from dancing through the reader's mind.

PROBLEMATIC PUNCTUATION

> *I spent all morning taking out a comma and all after-noon putting it back.*
>
> —Oscar Wilde (attributed)

Sometimes, it's not the words themselves but the marks put between them that can cause problems. Punctuation characters, such as the lowly comma, have frequently been at the heart of communication failures and disputes.

Before 1913, when a permanent income tax was instituted via the Sixteenth Amendment to the Constitution, the U.S. government was heavily dependent on tariffs for revenue. And an 1872 revision to the government's tariff act included what has been referred to as the most expensive typo in legislative history. A previous tariff act, passed in 1870, had specified a 10 to 20 percent import duty on fruits, and this constituted an important source of federal revenue. Other items, such as fruit plants, were exempted.

But in the act's revision, a comma was erroneously inserted between "fruit" and "plants" instead of a hyphen ("fruit, plants" instead of "fruit-plants"), which created the mother of all loopholes. Importers of fruit promptly filed for refunds on duties they had previously paid, but the Treasury Department refused to honor them, insisting that the comma had been a mistake. The importers took the government to court, which ruled in favor of the importers.

The federal government had to refund the duties it had collected, which came to about $2 million, or about $46 million in 2020 inflation-adjusted dollars.[18,19] And taking into account how much larger the federal budget has become, this single-character mistake is proportionately equivalent to tens of billions of dollars today.

The modern comma and its use can be traced to Aldus Manutius, an early Venetian printer who published the first editions of many early manuscripts. Before his innovation, a virgule (/) had often been used to denote a pause. He lowered this slash and curved it slightly.[20] The convention was quickly adopted by other publishers who recognized the need to separate the clauses of a sentence and to indicate separate items in a list. But a debate about when to employ this mark has a long history. Wrangling about the serial, or Oxford, comma has been particularly fraught, with partisans on both sides of the dispute sniping about clarity and consistency.

Devotees of the serial comma use it to separate three or more constituents of a list, including the item before a coordinating conjunction, such as "and," as in *red, white, and blue*. Punctuating a list in this way is common in American English and advocated for by most U.S. style guides. The serial comma is typically not used in British English, and this would change our example to *red, white and blue*. (It's called the Oxford comma because Oxford University Press—even though it is a British publisher—traditionally employed serial commas.)

While this distinction may seem like splitting hairs, serial comma advocates claim that using them prevents ambiguity. Mary Norris, who worked for many years as a copy editor for *The New Yorker*, provides a mentally indelible example in her book *Between You & Me: Confessions of a Comma Queen*:

We invited the strippers, JFK and Stalin.[21]

Without the serial comma, we are left to imagine the U.S. and Soviet leaders suggestively removing their clothing at a truly unusual social function. But by inserting the serial comma, Kennedy and Uncle Joe are simply part of a larger group that also includes striptease artists.

Many other imagined examples can be found on the internet, such as the grateful award recipient who exclaims, "I'd like to thank my parents, Bill Clinton and Oprah Winfrey." Without the comma before "and," the speaker is laying claim to a surprising ancestry. With the comma, the person is simply a shameless name-dropper.

Although it is an amusing exercise to create such sentences, there have been instances in which the failure to employ a serial comma has had real-world consequences. In 2014, truck drivers sued Oakhurst Dairy of Maine, claiming they had been deprived of pay to which they were entitled. They pointed to a state law that requires extra compensation for overtime work, with certain exceptions that were enumerated in the statute. Unfortunately for the company, the First Circuit Court of Appeals ruled that the lack of a serial comma created ambiguity about which activities were exempt from overtime payments. The dairy ended up paying the drivers $5 million in back wages.[22]

And for some, the interpretation of a comma in the Second Amendment to the Constitution makes its intended meaning problematic. As a reminder, its twenty-seven words state,

A well regulated Militia, being necessary to the security of a free State, the right of the people to keep and bear Arms, shall not be infringed.

In 2007, the U.S. Court of Appeals for the District of Columbia ruled against the district's handgun ban. In a ruling authored by Laurence

Silberman, he argued that the comma after "State" creates two clauses in the amendment and that, as a result, it not only protects the right of the states to maintain militias but also grants individuals the right to bear arms. This interpretation of the founding fathers' intentions has proven to be contentious.[23]

The semicolon is another punctuation mark that has caused grief for lawmakers and the courts. In her 2019 book about the semicolon, Cecelia Watson describes a number of legal cases that have hinged on how this mark was interpreted.[24] Even the lowly apostrophe has gotten in on the act. In October 2021, a missing apostrophe in a social media post (*employees* instead of *employees'*) was used as evidence that an Australian real estate agent intended to provide retirement funds for all of his workers. A ruling against the agent could cost him tens of thousands of dollars.[25] As all these examples suggest, punctuation can serve as a tool to reduce ambiguity and misunderstanding, but in at least some cases, it has had the opposite effect.

A NEED FOR SPEED?

There ain't no such thing as a free lunch.

—Origin unknown

We all know that when we do things quickly, we tend to make mistakes. A quick trip through the supermarket, for example, will almost inevitably lead to regret later when we realize we forgot about several items we were planning to purchase. Researchers refer to this as the speed–accuracy trade-off: when we do something slowly and carefully, we are unlikely to make many mistakes.[26] However, if we rush something, such as our shopping trip, the results are likely to be less than ideal.

We can do things fast, or we can do them well. But does this truism apply to language? Does reading too quickly or listening to rapid speech impair comprehension and understanding?

Let's consider the case of reading first. To begin, it should be noted that people read different types of texts for different purposes, and this greatly affects reading speed, comprehension, and later memory for what was read. For example, we might rapidly run our eyes down the middle of a list of names to see if we were chosen for some activity. This might allow us to quickly determine if we were selected, but we would probably have poor recall for the other names on the list at a later time. Or we might skim a news article with only enough attention to get the gist of the story.

In contrast, we might carefully pore over the description of a location in a whodunit on high alert for clues to the murderer's identity. A good ballpark estimate for the adult reading rate is about 280 words per minute (WPM).[27]

Can people learn to read faster than this? That is the siren song of those who advocate for so-called speed-reading: the tantalizing possibility that one's reading rate can be doubled or even tripled, all without sacrificing comprehension. However, research has consistently failed to support such claims: reading speed cannot be substantially increased without a commensurate decline in understanding.[28] Cognitive constraints seem to preclude the mental equivalent of a free lunch.

But could different ways of reading circumvent the speed–accuracy trade-off? Reading is a complex process that involves the eyes and brain in a bidirectional dance. One potentially limiting factor in this activity might be the necessity of moving one's eyes along lines of text. Perhaps it would be possible to eliminate the middleman by having the text move instead of the eyes. A technique called Rapid Serial Visual Presentation (RSVP) was created to accomplish this. Words are flashed, one at a time, in quick succession, and all the reader has to do is stare at the center of a screen.

RSVP was originally developed to study the reading process, but there are now apps that allow people to read in this way via their smartphones. By using these programs, readers can consume texts at rates that are much higher than normal reading speeds, but comprehension inevitably suffers.[29]

As it turns out, a reader's eye movements aren't an impediment but are, in reality, an important part of the comprehension process. Readers can skip over short, familiar words that can be inferred from context, and they can also tarry over unusual or unfamiliar words to make sense of them. RSVP techniques are insensitive to such factors: the words march by at the same lockstep pace. Stopping and going back to review something that was missed can be tedious. And, in general, reading this way is both fatiguing and frustrating.[30]

Let's turn to a consideration of comprehending spoken language. Estimates vary, but the normal speaking rate for English is between 150 and 190 WPM.[31] At the lower end of this range are radio news announcers, who speak more slowly to enunciate clearly and to make up for the lack of visual cues that can aid comprehension. At the higher end, adults engaged in spontaneous conversation may reach rates as high as 200 WPM.[32]

Keep in mind, however, that people typically read at rates that are much faster than normal speaking: about 280 versus about 170 WPM, respectively. Does this imply that people could comprehend language spoken at faster rates than is typical? Auctioneers, for example, may reach rates between 250 and 400 WPM. This feat is possible because they engage in highly formulaic and repetitive speech performances. Nonetheless, does this suggest that people might have unused processing capacity available for comprehending faster speech?[33] To put it another way, is speed *listening* possible?

This issue has come to the fore with the increasing popularity of audiobooks and podcasts. Some listening enthusiasts have gotten into the habit of playing these at rates significantly higher than the original recording, which is typically 150 to 160 WPM. Such time-compressed speech, as it is called, allows listeners to consume more of a podcast episode, lecture, or novel within a given period of time. Maybe there is such a thing as a free lunch after all!

Research suggests that speech rate can be increased by 25 percent without much loss in understanding, even for relatively complex information like instructions. However, when the rate is increased to 50 percent, a significant decline in comprehension is seen.[34] And it is worth noting that even though time-compressed speech *can* be understood at these sped-up rates, it may not be a particularly enjoyable experience. Raymond Pastore has found that research participants typically prefer to listen to speech with only about 10 percent compression, or about 165 WPM.[35]

To conclude our analogy, a lunch that is free may not be particularly appetizing. It seems that reading and listening at high rates of speed bumps up against cognitive constraints on our ability to comprehend language.

THE MALLEABILITY OF MEMORY

In December 2013, Nelson Mandela died at his home in Houghton, South Africa. The passing of the former statesman, at the age of 95, was marked around the world with tributes and words of praise for his accomplishments. Some people, however, claimed to remember that he had died in prison during the 1980s despite the fact that he served as the president of South Africa in the 1990s.

Instances of such collective misremembering are legion. For example, Curious George, the monkey in a series of children's books, has no tail, even though many people believe that he does. And Alexander Hamilton wasn't a U.S. president, even though a 2016 study found that 71 percent of the participants mistakenly believed he had served in that capacity.[36] Compilations of such misrememberings, involving the appearance of corporate logos, lines of movie dialogue, or the spelling of product names, are a regular staple of online clickbait and listicles.

But how can we explain such erroneous beliefs? Fiona Broome gave this phenomenon its name—the "Mandela effect"—when she realized she mistakenly believed Nelson Mandela had died in prison. She could even recall having seen news clips of his funeral and the mourning that followed his passing. She also discovered that other people shared her false memories for the event.

Broome has gone on to propose that such memories are evidence for the existence of parallel or alternate universes.[37] According to this way of thinking, such errors are akin to a glitch in *The Matrix* noted by characters in the 1999 movie. This explanation makes a bit more sense when one learns that Broome is a paranormal researcher and author. She has encouraged others to report such anomalies on the Mandela Effect website and has self-published a series of books that provide evidence for her thesis.[38]

Psychologists and other cognitive scientists have offered a more mundane explanation. A growing body of research suggests that such confabulations are common and reflect normal mental processes. Simply put, we tend to confuse things that have a high degree of conceptual overlap.

For example, research participants tend to mistakenly believe that the word "sleep" was presented during an experiment that included related words, such as "bed," "pillow," and "dream."[39] As it turns out, false memories can be created in the laboratory fairly easily. And such an explanation can help to explain the erroneous shared memories about Mandela's passing: some people may be confusing his demise with that of others, such as Steve Biko, a South African antiapartheid activist who did die in police custody in 1977.

A complementary explanation can be found in the classic research of Frederic Bartlett, a British psychologist of the early twentieth century. As you may recall from chapter 1, he demonstrated that people tend to "normalize" unusual objects or actions when asked to remember them. When college students were asked to recall a Native American tale, they inadvertently changed the unfamiliar into the familiar. Canoes, for example, were frequently misremembered as boats. And hunting for seals, which would have been a relatively exotic pastime for Bartlett's British subjects, was transmuted into the act of fishing.[40]

Let's consider how these psychological explanations of the Mandela effect fare in cases that are linguistic in nature. It has been claimed, for example, that many people mistakenly believe the HBO sitcom *Sex and the City* was titled *Sex IN the City*. As it happens, this error is relatively common in newspaper articles that refer to the series.[41]

Why might this be? A search of the 400-million-word Corpus of Historical English reveals that the phrase "in the city" is nearly nine times more common than "and the city."[42] In other words, as we read texts in English, we encounter the phrase "in the city" far more frequently than "and the city." And as a result, we may be unwittingly inclined to normalize the title of the TV series when we attempt to recall it.

Another oft-mentioned example of the Mandela effect concerns the Berenstain Bears. The children's book series featuring the anthropomorphic ursine family was created by Stan and Jan Berenstain in the 1960s. However, many people mistakenly believe that the name of the cartoon bear family is "BerenSTEIN."[43]

Once again, this may be explained by considering the relative familiarity of the two names. According to the website forebears.io, there are about 1,100 people worldwide who possess the "Berenstein" surname, making it relatively uncommon. "Berenestain," however, is even rarer than that: only about 30 people have this last name. In addition, there are many common German and Jewish surnames that end with "stein," such as Goldstein, Bernstein, and Weinstein.

According to 1990 U.S. census data for the 15,000 most common surnames, there are 20 that end with "stein" but only one that ends with "stain" (Chastain, if you're curious). It would seem, therefore, that many people unknowingly normalize the name of the bears to make it look like a more familiar or common surname.

And what are we to make of the fact that nearly three-quarters of the participants in the study mentioned earlier mistakenly believed that Alexander Hamilton served as a U.S. president?

The authors of the study report that their survey data were collected in May 2015.[44] A few months earlier, the wildly successful musical *Hamilton* had opened in New York City, albeit off Broadway. The cast includes three individuals who served as or became president (Washington, Jefferson, and Madison). In addition, Hamilton's visage appears on U.S. currency (the $10 bill), a distinction shared by five presidents (Washington, Jefferson, Lincoln, Jackson, and Grant) but only one other non-president (Franklin). Just as the words "bed," "pillow," and "dream" can make one believe they must have encountered "sleep," the company that Alexander Hamilton keeps causes him to be misremembered as a chief executive.

It would seem, therefore, that the Mandela effect can be explained as the result of mental activation caused by conceptual overlap and a process of normalization. The same psychological processes that govern memory in general can account for the creation of such false memories. In short, we probably don't need to posit the existence of parallel or alternate universes to make sense of why such confusions occur.

AT A LOSS FOR WORDS

On September 8, 2016, Gary Johnson appeared as a guest on *Morning Joe*, the popular MSNBC news program. The former governor of New Mexico had run as the Libertarian candidate for president in 2012 and came in third, capturing 1 percent of the popular vote. Nevertheless, he was running again and this time against two widely disliked major-party candidates, Hillary Clinton and Donald Trump. During the summer of 2016, Johnson had been polling as high as 13 percent.[45] This was only two points below the threshold needed to qualify for the upcoming presidential candidate debate. Perhaps his time had come.

One question from that morning's interview, however, would doom Johnson's quest for the presidency. It was asked by columnist Mike Barnicle. "What," he inquired, "would you do—if you were elected—about Aleppo?" This question led to the following exchange:

Johnson: [tentatively] About . . .

Barnicle: Aleppo.

Johnson: [blankly] And what is Aleppo? [glances quickly at the other journalists]

Barnicle: [incredulous] You're kidding.

Johnson: No!

Barnicle: [gravely] Aleppo is in Syria. It's the . . . it's the *epicenter* of the refugee crisis.

Johnson: [relieved] Okay, got it! Got it!

Barnicle: Okay.

Johnson: Well, with regard to Syria, um, I do think that it's a mess . . . [answer continues]

Later that month, Chris Matthews asked Johnson to name his favorite foreign leader. When he was unable to answer, he confessed, "I guess I'm having an Aleppo moment."[46] His attempt at humor only served to remind Matthews's audience of his earlier gaffe. Equating his mistake with having a "senior moment" did him no favors either.

Six weeks later, Johnson would again place third in a presidential election, garnering a bit more than 3 percent of the vote. His quixotic campaign, however, had essentially come to an end two months earlier on an MSNBC soundstage in Jupiter, Florida, as he struggled to understand what Aleppo referred to.

Not surprisingly, Johnson has been defensive when asked about his "Aleppo moment." However, in a 2018 interview with *Esquire*, he finally addressed the issue. "I forget my mother's name occasionally. We all do. I went in there and there was no context. So when he [Barnicle] said 'Aleppo,' I was thinking acronym. Wrong, but nonetheless."[47]

Johnson's explanation does have the ring of truth to it. His claim that "there was no context" might seem like a weak defense but deserves to be considered carefully. A web search will turn up multiple videos of Johnson's interview, but almost all of these begin with Mike Barnicle's question about Aleppo.

A full transcript of the interaction, however, shows that this query occurred near the beginning of the interview, after his introduction and a discussion of the two-party political system. Immediately before Barnicle's infamous question, he had been asking Johnson about the possibility that his candidacy would affect the presidential race in the same way that Ralph Nader's had in 2000.[48] Barnicle's next question, then, about a refugee crisis in Syria truly does seem like an unexpected pivot to a totally different topic.

In addition, Barnicle's query lacked a contextual frame: "What would you do—if you were elected—about ——?" Such a sentence could be the lead-in to virtually any topic: immigration, deficits, or the dangers posed by ravenous aliens.

Researchers who study language comprehension refer to such constructions as having low Cloze probability.[49] An educational psychologist might vary the Cloze probabilities of sentences, for example, to assess a child's linguistic abilities and vocabulary.

To understand this better, consider a variant of Barnicle's question that has a high Cloze probability: "What would you do—if you were elected—about the Syrian army's siege of ——?" In this case, the universe of possibilities is vastly reduced to cities under attack in the Middle East. And because of the way Barnicle chose to word his question, Johnson couldn't start making sense of it until the final word—Aleppo—had been spoken. And when Johnson requested clarification, Barnicle provided no assistance—he merely repeated the problematic word.

Finally, it seems unlikely that Johnson would have blanked if he had been asked about Alabama or Alaska, for example. Aleppo had been in the news a great deal during the summer of 2016 as Syrian government troops tightened their long siege of the city and the battle entered its decisive and final phase, creating a humanitarian crisis. Nonetheless, its name may have been exotic enough that, without context, it sounded like a stream of sounds, which could have suggested that it was an acronym.

Johnson was not the first presidential candidate to experience a mental lapse while on a national stage. In a November 2011 Republican debate, Texas Governor Rick Perry famously couldn't remember the name of one of the three government agencies that he had vowed to eliminate.[50] He sheepishly apologized for having stumped himself and finished with an "oops" (it was the Department of Energy, which he later headed during the Trump administration).

We normally associate word-finding problems with speaking and view them as an unwelcome harbinger of advancing age. And such failures can certainly have a negative impact on communication: the failure to produce a word on the tip of one's tongue can derail a conversation as both speaker and listener find their interaction devolving into a crude form of charades.

And research on this topic does suggest that word retrieval is affected by the aging process. In a representative study, a large group of Israeli adults were asked to name familiar objects, such as a hat, when confronted with simple line drawings of these objects. The average performance of participants increased until about age 50 and then declined steadily thereafter. (It should be noted, however, that there were large differences among the participants: Some subjects who were in their seventies performed as well as others who were a generation younger.[51])

Johnson was 63 at the time of his "Aleppo moment," and Perry was 61 when he forgot about the Department of Energy. And in both cases, the stakes were high for their performances. When we consider all the relevant factors, it seems likely that the circumstances created ideal conditions for language-related memory impairments.

8

SOCIAL FACTORS

L anguage is, at its heart, a social phenomenon. We use it to share our thoughts, feelings, and desires with others. We also use it to engage in social play. However, what one person might intend as playful banter may be interpreted quite differently by its recipient. In other cases, we might deliberately obfuscate our words so that they can't be understood by an overhearer. And even within the same nation, regional differences in pronunciation or word choice can affect communication.

JUST KIDDING

> *I can remember being bullied and teased. It was absolutely horrible. I got kicked out of ninth grade for throwing a book at a girl who teased me. It was absolutely terrible.*
>
> —Temple Grandin, *The Tavis Smiley Show* (2013)

A critical aspect of successful communication involves assessing the intentions of one's conversational partner. For example, is she being serious or playful? Nonserious banter occurs frequently in brief and casual conversations, such as when people engage in small talk or congregate around the watercooler at the office.

The intent of such nonserious talk, however, can be ambiguous, which makes it a problematic form of speech. A remark could be interpreted as good-natured ribbing, but it might also be perceived as commiseration, flirtation, veiled criticism, or something else. It is the inherent ambiguity of such language that can count as a strike against clear and transparent communication.

One particularly troublesome form of nonserious talk is teasing since it can be interpreted positively, perhaps as lighthearted repartee, or negatively, possibly as harassment or bullying.[1] And there's nothing worse than the joke that isn't understood as a joke.

As we've already seen, an important aspect of inferring communicative intent involves successfully reading someone's body language, tone of voice, and relevant situational cues. And when one or more of these is absent, as when communicating by phone, email, or a text message, determining intent can be even more difficult. For now, let's consider face-to-face communication in the context of teasing, which often involves deliberate provocations, such as unflattering observations about a person's appearance, characteristics, or behavior.[2]

Clearly, this can be a risky form of communication. Teasing can be described as a face-threatening act, as it places in jeopardy an individual's desire for social affirmation from others. However, from a societal perspective, such teasing may play the important role of enforcing group norms and expectations. Parents, for example, engage in such talk to gently discipline or instruct their children about acceptable conduct.[3]

From a developmental perspective, an understanding of the nuanced nature of teasing emerges only gradually. Younger children, for example, conceptualize teasing as a purely negative phenomenon. However, by middle school, they also engage in such talk with their friends and make use of context to disambiguate such remarks. High school students are capable of using teasing to strengthen social bonds, to deal with embarrassment, or even to bring up problematic topics that they don't feel comfortable discussing directly.[4]

Researchers who study teasing have noted a disconnect between the intentions of teasers and the recipients, or targets, of such remarks. Although the teaser will often attempt to reduce the negativity of what he has said through nonverbal and other means, his target may not notice these accompaniments or may be insufficiently mollified by them. As a result, teasers typically view their banter in a relatively benign light. However, when the tables are turned and people reflect on their experience of being teased, they view such talk more negatively.[5]

If it appears that the target is taking offense, the person doing the teasing may attempt to block or even forestall such a reaction by adding "Just kidding" or asking "Can't you take a joke?"[6] Such amendments can put the recipient of teasing in a difficult position; her feelings may have genuinely been hurt, and then insult is added to injury when she is informed that she lacks a sense of humor. It's not hard to imagine such episodes escalating to verbal aggression.

Teasing has mostly been studied in a small number of specific contexts, such as in close relationships, the classroom, and the workplace. For example, a study of workplace humor in Australia found that teasing is a commonly employed tactic to build rapport or as a prod to get things done.[7]

And a critical factor that affects how teasing is perceived is the closeness of the relationship between the teaser and her target. In general, teasing is perceived more positively within close friendships and romantic relationships than it is between people who lack such ties.[8]

Teasing often involves the use of sarcasm, in which ostensibly positive statements, such as "You're a real genius," are intended negatively. This type of teasing is particularly useful because it allows people to express negative feelings in socially acceptable ways.[9]

Jokes can be problematic because it may not be obvious when someone is joking. This is particularly true for online posts. In January 2020, a teacher at Babson College was fired for his response to President Trump's threat to bomb cultural sites in Iran. The adjunct professor posted a list of sites on Facebook that he said Iran should target—such as the Mall of America and a home of one of the Kardashians. The post circulated widely on social media and predictably sparked online outrage.

The instructor protested that this was only a joke he had made with his friends. And even though others described his post as "obvious rhetorical hyperbole," he still lost his position.[10] Because such episodes can be damaging to an institution's reputation, the leaders of universities and other high-profile organizations often feel they have no choice but to respond forcefully to such failed attempts at humor. The risks posed by potential misunderstandings seem to trump freedom of speech in such cases.

HAVE YOU HEARD?

> *A lie can travel halfway around the world while the truth is putting on its shoes.*

This quotation has been attributed to a number of individuals, including Winston Churchill, Mark Twain, Thomas Jefferson, and Ann Landers. There is no evidence, however, that any of them ever said it.[11]

The lack of a source, however, didn't stop the *New York Times* from attributing the quote to Twain in stories published by the paper in 2008 and 2010.[12,13] (Ironically, the 2010 piece was an editorial by Thomas Friedman about the importance of fact checking.) Twenty years earlier, Massachusetts Governor Michael Dukakis had also employed the line, as well as the attribution to Twain, in a critique of his opponent, Vice President George H. W. Bush, during the 1988 presidential campaign.[14]

These misstatements weren't lies: it's not as if Friedman or Dukakis knew the true author of the quote but chose to credit the famous humorist instead. Passing along information because it is plausible or believed to be true goes by many names, including folklore, old wives' tales, superstition, urban legends, scuttlebutt, rumor, gossip, and hearsay. These terms aren't synonymous—the misattribution of a quote isn't the same as a child reciting "Step on a crack, break your mother's back"—but they are all labels for different facets of the same phenomenon: communication that facilitates the spread of misinformation.

As we know all too well, the internet has made it much easier for factually incorrect or poorly sourced information to travel "halfway around the world." Online forums and social networking sites have greatly increased the odds of encountering information of questionable legitimacy, such as conspiracy theories, predictions about the end of the world, or other hoaxes.

And it takes only a few swipes or keystrokes to retweet or "like" a post, forward an email, or broadcast copypasta. At the darker end of such activities, the online world allows people to engage in cyberbullying or the doxing of their enemies. But while the transmission of gossip and rumors has become easier, these phenomena have undoubtedly existed for as long as people have been using language.

Research has shown that people tend to use the terms "gossip" and "rumor" interchangeably. Scholars, however, have argued that an important distinction can be drawn between them: rumors are "unverified and potentially useful information." They come into being in ambiguous or threatening situations, and they can be useful in understanding and dealing with some threat. Gossip, on the other hand, refers to "evaluative social chat about individuals" and fulfills a variety of functions, such as increasing one's own status or maintaining social norms.[15]

A 1998 study of gossip on a college campus found that, in two-thirds of such cases, the information was intended to embarrass or shame someone.[16] But gossip may not be completely deserving of its bad reputation. It can play a positive role within groups since it allows important information to be transmitted widely and efficiently. It can also create and strengthen the bonds that exist between group members.[17] Seen in this way, casual watercooler conversation should be thought of not as idle chitchat but as a source of relationship building and social cohesion in the workplace.

On the other hand, those who start or spread gossip are typically perceived negatively. They might be labeled as troublemakers or pot stirrers. And despite the widespread stereotype, studies have repeatedly demonstrated that men gossip as much as women.[18]

Should gossip be considered a form of miscommunication? The answer seems to depend on how faithfully such information is transmitted. For most of human history, gossip traveled from person to person by word of mouth. And as described in chapter 7, children can learn about the fragility of such informational chains by playing the Telephone game.

Just as a copy of a copy of a copy tends to become distorted and corrupted, a snarky remark made by a jealous coworker can transmogrify into a vicious attack. As L. M. Montgomery, the author of *Anne of Green Gables*, put it, gossip is "one-third right and two-thirds wrong."[19]

The scientific study of rumor began with the work of psychologists Robert Knapp, Gordon Allport, and Leo Postman during World War II, a time of great uncertainty for many Americans. Among other things, these researchers proposed that the strength of a rumor is related to its importance and degree of ambiguity that exists about a particular topic.[20] In other words, factors like personal relevance can affect whether a rumor is promulgated. If there is an information vacuum, rumor may rush in to take its place. This was certainly true during the war, when many families had loved ones serving abroad and the government withheld information like troop movements from the public. And as with gossip, it seems that rumors are susceptible to the distorting echo-chamber effect caused by long chains of repetition.

Hearsay, in contrast, is a legal concept and refers to the reporting of someone else's words or behavior by a witness in court. If the other party is not available for cross-examination, such a report is inadmissible as evidence. In the United States, this is dictated by the Sixth Amendment to the Constitution, which provides those accused of a crime the right to confront their accuser.

As it happens, there are a number of exceptions to the rule against hearsay. For example, an attorney can make use of a "learned treatise," such as an article published in a medical journal, to rebut the testimony of an expert witness.[21] The courtroom, therefore, is one venue in our public lives that allows reliable information to counter misinformation.

THEY DO THINGS DIFFERENTLY THERE

It was John Donne who asserted, "No man is an Island, entire of itself." And just as no one is an island, no language is a monolithic system. With a language like English, a quotation from Walt Whitman may be more appropriate: "I am large, I contain multitudes." One way in which these multitudes manifest themselves is through different dialects of the language. And they constitute yet another potential vector for miscommunication and misunderstanding.

Let's first consider such variation at a national level. In the United States, there are reasonably well-defined regional dialects in the eastern part of the country. (For reasons having to do with settlement patterns, the country becomes more homogeneous linguistically as one moves farther west from the areas comprising the nation's original colonies.) When most people hear the term "dialect," they think of variations in pronunciation, or accents, although as we will see, there are other regional differences as well.

Pronunciation differences and the misunderstandings they cause are sometimes exploited for comic effect on television and in the movies. A good example can be found in the 1992 comedy *My Cousin Vinny*. A lawyer from Brooklyn, Vinny Gambini (played by Joe Pesci), is representing his cousin in an Alabama murder trial.

At one point, Vinny says, "Is it possible the two utes," and is promptly interrupted by the courtly southern judge (Fred Gwynne), who asks, "Two what? What was that word?" After some back-and-forth, Vinny finally understands the judge's confusion and carefully enunciates the word "youths" with a hyperarticulated "th" sound. This characteristic of the New York City dialect, referred to as "th-stopping," is one of a number of distinctive features of this accent.[22]

In real life, however, such confusions may not be especially common. Consider a Bostonian who might (stereotypically) utter a phrase like "Pahk the cah in Hahvahd yahd." Although a tourist from Seattle or Dallas might find the accent in Beantown to be somewhat unusual, it would probably present few difficulties in terms of comprehension.

There are, however, lexical and syntactic differences between dialects, and these can be more consequential. English spoken in southern Louisiana, for example, was influenced by French-speaking Acadians who migrated there from Canada's maritime provinces in the eighteenth century. As a result, Cajun English has a number of words derived from French that are not found in other dialects of the language. "Cher" (dear), "mange" (to eat), and "fache" (angry) are a few of the terms that might trip up the linguistically unwary.

Other lexical differences manifest themselves at a national level. If you were to decide to quench your thirst with a sweetened carbonated beverage, for example, you might say, "I think I'll have a ——." The term that someone might use to complete the sentence will be heavily influenced by the dialect of American English they have been exposed to. Broadly speaking, people living in the northeastern United States and California would most likely say "soda," whereas those living in the Midwest and the northern tier of the country would opt for a "pop."[23] There are some exceptions—the St. Louis and Milwaukee areas, for some reason, are islands of "soda" in a sea of "pop"—but in general, these patterns are remarkably consistent.[24]

But what about the American South? That's were things become a bit complicated. The default term in a region stretching from South Carolina through Texas is "coke," which might seem confusing if you are not a southerner. After all, isn't "coke" just one type of soda pop?

The explanation seems to be tied to regional identity: Coca-Cola was created by pharmacist and Civil War veteran John Pemberton in 1886. A war injury led to a morphine addiction and experiments with analgesic alternatives, such as a concoction that included the extract of coca leaves.[25] This patent medicine evolved into Coca-Cola, the first mass-produced and widely available soft drink. It was initially manufactured and bottled in Atlanta, the city that still serves as its worldwide headquarters.

And if we consider other varieties of English, such lexical differences become even more pronounced. George Bernard Shaw, the Irish playwright, allegedly referred to the United States and Great Britain as two countries separated by a common language. Regardless of who said it, the speaker was undoubtedly thinking about the many words that mean different things in the two varieties of English, such as "bog" (swamp versus toilet), braces (teeth straighteners versus suspenders), and trolley (shopping cart versus streetcar).

There are differences between the United States and other forms of English as well. A Canadian might wear a "tuqe" instead of a knitted cap, live in a "bachelor apartment" instead of a studio apartment, and leave her car in a "parkade" instead of a parking garage.

Australian English is well known for its inventive and colorful vocabulary, such as "fair dinkum" (real or true) and "bogan" (a provincial person). A "grommet" isn't an eyelet but a young or novice surfer. "Root" is slang for "to have sex with," an act that might start with a "pash" (a long and ardent kiss).

At the syntactic level, we find peculiarities in the American South that have nothing to do with beverages. Some southerners make use of so-called double- or multiple-modal constructions, such as "might can," "might could," and even "might should ought."[26] Although this construction has not received extensive study, it seems to occur most frequently as a form of indirect speech in the context of negotiation.

Functionally, these constructions seem to be employed as a form of face-saving. Multiple modals are often used in directives ("You might ought to have the oil changed") and conditionals ("We might shouldn't have done that last night").[27] The construction appears to have migrated with the Scotch-Irish to America in the eighteenth century. Speakers of other dialects of English who are unfamiliar with such constructions might perceive this politeness strategy as a sign of equivocation or a lack of education.

FOXY DOCTOR FOX

It was the summer of 1970, and the attendees at a faculty retreat near Lake Tahoe settled in to hear a lecture on mathematical game theory and its application to physician education. The audience consisted of 11 psychiatrists, psychologists, and social work educators. The speaker was introduced as Doctor Myron L. Fox. In introducing him, Fox's host enumerated his impressive academic credentials and publications, making it clear that he was an expert on the topic of his speech.

Throughout his 40-minute address, Fox was charismatic and dynamic. His disquisition, however, was full of double-talk, platitudes, contradictions, and pointless digressions. But it wasn't Dr. Fox's fault: he didn't know much about game theory. In fact, he didn't even exist.

Doctor Myron Fox was in reality Michael Fox, a professional character actor. He had often *played* the role of a physician on television programs like *Perry Mason* and *The Twilight Zone*, but that was the extent of his medical knowledge. He was provided with only a single article on game theory and was given one day to practice before giving his address. His remarks come across like a stream-of-consciousness daydream, never truly connecting with the stated purpose of his speech.

And how did the assembled physicians and educators react to his lecture? On the whole, responses were favorable: they gave him high marks on a questionnaire assessing their satisfaction with the presentation. It appeared that the professionals who composed his audience were more interested in being entertained than informed.

The researchers concluded that those who heard Fox speak came away with an illusory belief that they had truly learned something. To check their findings, the researchers showed a recording of Fox's lecture to two additional groups of well-educated professionals and obtained similar results.[28]

The study's conclusions, known forever after as the Doctor Fox effect, were criticized on methodological grounds, such as the fact that it lacked a control group. To counter this objection, the researchers conducted a follow-up study with 200 college students. Actor Michael Fox reprised his role as Doctor Fox and gave lectures on the biochemistry of memory.

Half of the lectures were delivered in the same seductive style that Fox had used with the professionals at Lake Tahoe. His delivery was heavy on humor, enthusiasm, and charm but also larded with non sequiturs, irrelevant content, and parenthetical asides. In another set of lectures, his remarks were deliberately less engaging and entertaining but contained the same content.

Afterward, all the students completed a satisfaction questionnaire and a test on material from the lecture. Those who had heard Fox speak in an amusing and compelling way provided higher satisfaction scores regardless of how much actual content the lectures contained.[29]

The Doctor Fox studies have often been cited as an important critique of the validity of student evaluations of their instructors. The claim is that students provide high evaluations to teachers they find to be entertaining as opposed to informative or even comprehensible. Style, it would seem, triumphs over substance.

But the real story may be a bit more complex than that. In 2014, an exact replication of the original study, using a video of the lecture given in 1970, found the same Doctor Fox effect as in previous experiments. In one version of this replication, however, the researchers explicitly asked participants if they had learned anything from the lecture. The participants reported they had not.

In other words, there was no evidence for illusory learning; instead, it simply appears that people enjoy speakers who are entertaining. The mistake made by the original researchers was to implicitly equate the act of enjoyment with evidence of learning.[30]

A parallel line of research on persuasion has found that the expertise and attractiveness of a communicator can have effects that outweigh the message itself. When listening to someone with our full attention, we tend to be swayed by relevant factors, such as the quality of the speaker's arguments. However, this assumes that we have both the ability and the motivation to attend to the message carefully.

In many situations, however, we don't pay careful attention to the message or simply aren't that interested. In these cases, our attitudes may be swayed by irrelevant factors, such as things that make us feel good or how entertaining we find something to be.[31] It's why television commercials tend to employ famous people, pets, babies, or other props that people enjoy looking at.

For the advertiser, creating a positive link between a potential consumer and a product may be good enough. In the future, these positive associations may be sufficient to influence consumers' opinions or, ideally, their purchasing behavior.

For our purposes, the Doctor Fox effect provides a compelling example of how easily the attractiveness of a communicator can override the content of his message. This seems to be especially true when someone is billed as an expert within a given domain. Even if his words don't make a great deal of sense to us, we can make a variety of attributions to explain away the non sequiturs and double-talk that someone like Doctor Fox might employ. For example, we might conclude that we're not smart enough or lack the expertise to truly understand what the speaker is talking about. Or even if we notice the discrepancies and contradictions, we might make an external attribution: the speaker is having a bad night or simply got mixed up while speaking.

A good example of this occurred in May 2008 during the Democratic primary campaign, when then-Senator Barack Obama claimed to have visited 57 states. From the context of his remarks, it's fairly clear that he meant to say, "47 states."

Pundits who supported Obama chalked it up to fatigue or a slip of the tongue. Extreme voices from the right, however, proclaimed that this was a reference to the 57 member states in the Organisation of Islamic

Cooperation and further proof of the president's Muslim leanings.[32] Clearly, our positive or negative feelings toward a speaker can play a major role in how we interpret what they say.

MASKING ONE'S MEANING

There are times when we want to communicate with someone but don't want our words to be known by others. If our conversational partner is physically present, we can simply make the message inaudible to others by whispering into his ear. Depending on our relationship with this person, however, such a strategy could range from merely awkward to highly inappropriate. It also broadcasts to everyone else that a private conversation is being conducted, which is rude behavior at best. Switching to another language that is known by you and your partner but not by others present may be perceived as impolite as well.

Other forms of concealment are more involved. Imagine a couple discussing a potential outing in front of their young child. They might, for example, find themselves spelling words out as they discuss possibly going to the "z-o-o." Similarly, they could resort to pig latin and refer to an "iptray to the oozay." Such dodges, however, will work only if the child is too young to spell or is unfamiliar with more complex ruses.

Another alternative available to people who know each other well is to create an idiosyncratic code on the fly, perhaps by referring to shared experiences that only they and their partner know about. These types of obfuscation would not count as deception since the speaker isn't attempting to confuse or to lie to her conversational partner. However, a delicate balance must be maintained between insufficient concealment on the one hand and creating messages that are too obscure on the other.

Research suggests that people are relatively adept at getting this balance about right. Psychologists Herb Clark and Edward Schaefer conducted an experiment in which the study participants were pairs of friends at Stanford University. They were asked to arrange a series of pictures of familiar campus scenes and landmarks into a particular order. The friends could hear but not see each other or their partner's pictures.

One student in each pair was the director: she was assigned the task of describing the pictures to her friend, the matcher, as best she could.

During some trials, however, the director was encouraged to actively conceal her descriptions from an overhearer who was performing the same task as her friend. During these concealment trials, the director spontaneously referred to the pictures in unusual ways that could not be understood by the overhearer. For example, instead of simply referring to locations by name, such as Hoover Tower or Stanford Stadium, the director said things like "involves my summer job" or "the closest place to where I live right now."

Clark and Shaefer referred to such idiosyncratic utterances with a term borrowed from cryptography: they were "private keys" that allowed their friends but not an overhearer to crack the code.[33] And such codes rely on the shared common ground between two people, as we saw in chapter 2.

In another study, researchers investigated something more involved than mere concealment. There are times when we wish to broadcast two meanings at once: one that will be interpreted in a certain way by some and in a different way by others. This is referred to as the multiple-audience problem. To study this phenomenon, researchers asked participants to conceal a piece of information in a written or videotaped essay. Specifically, they were told to craft their message in such a way that the concealed information would be decipherable by their friends but not by strangers. The participants found the task difficult, but they often stumbled on the strategy of writing or saying things their friends would know to be untrue.

In one version of the experiment, for example, the participants' task was to communicate the identity of a song from a set of four alternatives. One subject accomplished this by stating that he found a particular song's religious symbolism to be annoying and went on to claim that religious messages have no place in popular music. This participant happened to be deeply religious, and his friends used his deceptive statement to help them ferret out the correct choice.[34]

Telling transparent falsehoods is one way to solve the multiple-audience problem, but people can make use of other strategies as well. Verbal irony can also function in this way. Its use can allow members of one's inner circle to divine a meaning that is distinctly different from the literal interpretation of one's words. An overhearer may lack the knowledge or context to arrive at this alternative meaning.[35]

But as we saw in chapter 7, there is no such thing as a free lunch. The results of the experiments described above make clear that using private keys or creating transparent falsehoods is a cognitively demanding activity. Typically, the research participants took longer and became more dysfluent when they had to craft such indirect statements. In addition, these attempts at concealment led to higher error rates.

The pairs of friends at Stanford, for example, were almost perfect in arranging the campus pictures when they could refer to them in any way they wanted. When the director was asked to engage in impromptu message masking, however, the matchers' performance, although still high, dropped significantly. Concealment, in other words, increased miscommunication.

Another complication is that the directors' attempts at concealment weren't all that successful. In the study involving the use of private keys, an overhearer was given the task of trying to arrange the pictures in the same order as the director and matcher. As might be expected, the overhearer did worse at this task than the pairs who were friends. The overhearers' overall level of performance, however, was significantly above chance.

This result makes sense since the clues that a director could use to describe palm trees in the central quad would probably be different from how she might describe the Memorial Church. The overhearer could make use of these differences to infer whether the director was describing a stand of trees or a building.

It could be argued, therefore, that masking one's meaning carries with it the risk of insufficient concealment at one extreme and miscommunication at the other. Truly cryptic messages may be as likely to bewilder a young child or a nosy neighbor as to confuse the intended recipient of one's words. Miscommunication of this sort can be described as the carefully manufactured but unintended result of such over-concealment.

DELIBERATE MISUNDERSTANDING

What we have considered thus far are examples of unintentional misunderstanding, that is, cases in which a listener or reader attempts to divine what someone means but is unsuccessful for some reason. However, it is also the case that someone could deliberately *act* as if they misunderstand to achieve a particular goal.

For example, as we will see in chapter 10, a hostile witness may deliberately misinterpret the questions of a prosecutor to protect the accused—or themselves—from being implicated in criminal activity. It might be a worthwhile exercise to consider such deliberate acts to see how they differ from misunderstanding arising from the factors described in the earlier sections of this book.

A good example of deliberate misunderstanding has come to be called the Trollope ploy. It was famously made use of by President Kennedy during the Cuban missile crisis in October 1962. Toward the end of that 13-day standoff, the U.S. government received a letter from Nikita Khrushchev, the Soviet leader. He promised to remove ballistic missiles from Cuba if Kennedy promised not to invade the island. The following day, however, Khrushchev publicly stated that he would remove the missiles only if the United States dismantled its Jupiter missile installations in Turkey. Clearly, the first offer was better than the second, but how should the United States respond?

It was Bobby Kennedy, the president's attorney general, who proposed a solution. He recalled a plot device from the writings of Anthony Trollope in which a woman willfully misinterprets the squeezing of her hand as a proposal of marriage. The attorney general suggested that his brother simply ignore Khrushchev's later public statement and respond only to the initial offer in the letter. The acceptance of this deal by the United States ultimately ended the crisis.

Although this version of events was not entirely accurate—Kennedy secretly promised Khrushchev that he would also remove the Jupiter missiles in Turkey—these details became public only long after the fact. By then, Kennedy's Trollope ploy had become firmly entrenched in the lore surrounding the crisis.[36]

In one sense, deliberate misunderstanding can be thought of as a violation of what Paul Grice called the cooperative principle.[37] This doctrine presupposes that both parties will make a good-faith effort to communicate successfully. When our conversational partner isn't willing, for whatever reason, to put in the work necessary to understand us, we might accuse them of being deliberately obtuse. This can happen, for example, if someone is seeking to duck responsibility for something they have done or to avoid something they are being asked to do.

Deliberate misunderstanding as a strategic act can be found in a variety of venues, such as the world of retail sales. This customer service tactic is sometimes referred to as "killing with kindness." An employee, such as a server or cashier, simply ignores the rude or provocative remarks that an irate customer is shouting at them. Instead, they respond politely and respectfully and seek to resolve the customer's complaint. The idea is to mollify or disarm the angry patron so that a potential sale or future relationship isn't jeopardized. And for those whose income is greatly affected by customer tipping, this skill is truly essential.[38]

As a strategy, killing with kindness can work fairly well—the calm demeanor of the employee may cause the patron to realize that he is being unreasonable. In other cases, however, this approach may be ineffective, and it can even backfire if the customer's goal is to upset the employee.

Another form of deliberate misunderstanding involves instruction and pedagogy. In Plato's dialogues, Socrates was frequently depicted as pretending that he did not know something or that he did not understand what someone was telling him. He did this to draw out the implicit assumptions and beliefs of his interlocutor, thereby exposing flaws in his reasoning. This technique, which is often referred to as Socratic irony or the Socratic method, was once commonly employed in legal education. Its popularity as a pedagogical device, however, seems to have waned.[39]

Is there such a thing as deliberate mis*hearing*? In some situations, a person may pretend not to hear something because it would be disadvantageous for them to do so. It is easy to imagine that a spouse's query, such as "Do you still have your boots on?" might be followed by a request to take out the garbage. This might cause her partner, in an attempt to avoid a cold trip to the curb, to pretend not to hear his spouse's question. (This would also buy him some time to remove his footwear.)

In other cases, a person may use headphones or earplugs as props to ostensibly demonstrate an inability to hear even though they may still be able to perceive and understand the words of others. The discussion of Mondegreens in chapter 3 presupposes that the mishearing of a song lyric is accidental. However, it is also that case that someone might pretend to mishear or misinterpret a snatch of a song for comedic effect.

Deliberate mis*reading* also seems to occur. An attorney, for example, may choose to interpret a statute more narrowly or more literally than its authors intended. Jill Anderson has pointed out that regulations involving "opaque" verbs (such as "desire," "promise," "intend," or "believe") are much more difficult to interpret than those involving "transparent" verbs (such as "touch," "borrow," or "send").[40]

The ambiguity of such terminology may create legal loopholes that can subvert the intentions of the legislative body that drafted the statute. Deliberate misreading has also been discussed in the context of literary criticism. Harold Bloom, for example, is well known for his theory that poets and other writers intentionally (and creatively) misread the work of earlier authors so that they can discover their own themes and voices.[41]

What these disparate cases have in common is the intentional *performance* of misunderstanding as a deliberate and strategic act. One could argue, therefore, that these are deviant cases and should not be considered genuine instances of miscommunication. Pretense is not the same as deception, but the intentionality that lies behind such behavior clearly separates it from the cooperative spirit that normally obtains in legitimate acts of communication.

9

CONTEXTS, PART 1

In the final two chapters of this book, we explore miscommunication in specific contexts. This discussion will build on issues that we have explored in previous chapters but also take into account the peculiarities of particular circumstances or forms of communication. What, for example, explains the unusual terminology employed by business executives? Why do authors frequently claim that people misunderstand what they have written? How does the process of translation affect the way that words, sentences, and even entire texts are understood? What role does the media play in creating misunderstanding? And why is flirting so often misinterpreted?

OUT OF CONTEXT

On December 20, 2013, on what would become the worst day of her life, Justine Sacco boarded a flight bound for London. The 30-year-old woman was the director of corporate communications for InterActiveCorp (IAC), a media company in New York City. She was looking forward to spending the Christmas holiday with her extended family in South Africa, where she had been born.

Once she arrived at Heathrow, Sacco passed the time before her next flight by making a few snarky posts on Twitter, such as one about a fellow passenger on the previous flight who had a body odor problem and another about the sorry state of British dentistry. Shortly before boarding her 11-hour flight to Cape Town, she tweeted,

Going to Africa. Hope I don't get AIDS. Just kidding. I'm white!

At the time, Sacco had 173 followers on Twitter. One of them chose to forward her tweet to Sam Biddle, the editor of *Valleywag*. Biddle, who had a reputation for being a provocateur, retweeted Sacco's post to his 15,000 followers and ran it online under the title "And Now, a Funny Holiday Joke from IAC's PR Boss." Soon the online world was atwitter about Sacco's outrageous post. A theory that her account had been hacked was discarded when other politically incorrect tweets she had made were unearthed.

The hashtags #JustineSacco and #HasJustineLandedYet began to trend worldwide, all without Sacco's knowledge since she was still en route to her destination and without internet access. When she arrived in Cape Town and discovered what had happened, she had a friend delete the tweet, along with her entire account. The following day, she was fired by IAC. Sacco apologized for her tweet in *The Star*, a South African newspaper, and her mea culpa reached a larger audience via ABC News.

All of us say and write things to friends and intimates that are not meant for public consumption, perhaps because they are intemperate, impolitic, or not intended literally. Given that Sacco had a relatively small number of Twitter followers, she probably felt that her post would be read in the spirit that she had intended—as caustic and edgy, perhaps, but not as evil or ignorant. That certainly fit the persona she projected online: her Twitter profile included the descriptive phrases "Trouble-maker on the side" and "Known for my loud laugh."

It probably didn't help that Biddle's repost mentioned that Sacco was a "PR boss" at a large media firm or that it was the holiday season, a time of the year when people ostensibly display more kindness to those who are less fortunate. These factors conspired to underscore a perception that the tweet was the clueless babbling of a privileged white woman who didn't know what she was talking about.

As it turns out, a name had recently been given to this phenomenon. In 2011, Alice Marwick and danah boyd pointed out that social media flattens multiple audiences into a single entity. They referred to this as "context collapse,"[1] and it explains why it is difficult to craft a social media post that will be understood in the same way by all of its recipients.

Think about how hard it is to create a Facebook update when you know that your young children, elderly grandparents, friends, and boss will all see it. In such situations, people tend to produce posts that are fairly bland so as not to offend anyone. Research has demonstrated that when an imagined audience is large, people tend to play it safe.[2] On the other hand, when posts are intended for only a small circle of friends, a person can be as outrageous or as shocking as they care to be. But when strangers get hold of and begin to parse such posts, a different dynamic comes into play.

Justine Sacco was an early victim of a phenomenon variously referred to as online shaming or cancel culture. The targets of such campaigns are usually well-known individuals, such as Roseanne Barr, who made a derogatory remark about Valerie Jarrett on Twitter and lost her television series as a result.[3] Another example is Kevin Hart, whose homophobic tweets cost him the hosting duties at the 2019 Academy Awards.[4]

But social media posts have also caused problems for lesser-known individuals. Kathy Zhu, who had been crowned Miss Michigan in 2019, lost her title when problematic tweets of hers came to light.[5] Ken Bone, along with his soon-to-be-famous red sweater, enjoyed 15 minutes of fame as a questioner at a 2016 presidential debate. He was later criticized for unsavory posts he had made to forums on Reddit.[6]

With the exception of Barr and Hart, none of these people intended their politically incorrect comments to be seen by a large audience. And when their later celebrity caused others to excavate their previous, semi-private musings, the backlash has often been swift and severe—just as it was for Sacco.

Sacco's own account of the episode made her intentions clear. In an interview with Jon Ronson, she said, "I thought there was no way that anyone could possibly think it [the tweet] was literal."[7] And a few months after the episode, Sacco reached out to Sam Biddle, the person directly responsible for fanning the flames about her tweet. In an article about their meeting for *Gawker*, he wrote,

Her tweet was supposed to mimic—and mock—what an actual racist, ignorant person would say. Ergo, tweeting that thought would be an ironic statement, a joke, the opposite of what it seemed to say. Not knowing anything about her, I had taken its cluelessness at face value, and hundreds of thousands of people had done the same—instantly hating her because it's easy and thrilling to hate a stranger online.[8]

"Not knowing anything about" a person but nonetheless taking her statement "at face value" seems like a perfect description of context collapse. Social media has made communicating with large numbers of people easier than it has ever been. Unfortunately, it has done the same for miscommunication as well.

GARBAGE LANGUAGE?

Uncanny Valley, a memoir by Anna Wiener, is an entertaining account of the author's career in Silicon Valley tech support during the late 2010s. Throughout her narrative, she lampoons a variety of Left Coast tropes associated with the technology industry. Among the many peccadillos she describes is the vocabulary employed by start-up entrepreneurs, influencers, and so-called thought leaders.

Specifically, Wiener objects to the "inscrutable jargon" of terms like "leading-edge solutions" and "first-mover advantage," characterizing such expressions as "garbage language." And her critique is only one of many that have been made throughout time by those who see the business world as running roughshod over plain speaking and understanding.[9]

"Buzzword," a post–World War II coinage, is a term that has long been associated with the peculiar lingo employed in large companies and bureaucracies. Similar terms include "business speak," "corporate jargon," and "management speak."

Concerns about such language have appeared in the media on a regular basis. The author of a 1973 article on jargon in the *New York Times* was particularly scathing with regard to the terminology employed in business management and advertising. Among other terms, "management by objectives," "reward factor," and "image identification" were characterized as "idiot talk."[10]

Fourteen years later, an editorial, also appearing in the *Times*, described jargon as "the growth industry of the 80's."[11] A 1999 *Times* article decried the use of terms like "administriva," "elephant hunt," and "dead cat bounce."[12] And yet another columnist, writing for the *Times* 16 years after that, took umbrage with terms like "bandwidth," "deliverables," and "deep dives."[13] The authors of these articles, published during a 40-year period, come to remarkably similar conclusions: jargon is bad, people use too much of it, and enough is enough.

Why do the buzzwords employed by corporations have such a bad reputation? One reason may be that they are often used euphemistically. As we saw in chapter 4, euphemisms are employed by other professionals, such as physicians and the military, to refer obliquely to unpleasant topics like mortality and death. And as with doctors and generals, the denizens of corporate boardrooms often have to make relatively cold-blooded decisions that can have far-reaching and negative consequences for others. As a result, their pronouncements are often cloaked in circumlocution.

Consider the language of termination. At one time, employees who were fired were described as having been "let go"—as if the employee had been given permission to break free from the warm embrace of her employer instead of losing her job. Today, a corporation is more likely to refer to "streamlining," "downsizing," "reductions in force," or even "right sizing" to describe such dismissals. After all, who could object to the creation of "operational efficiencies"? The business world has been enthusiastic in coining euphemistic expressions for these and other less-than-savory corporate acts. It's reached the point that employees play "buzzword bingo" at meetings, using cards replete with the jargon employed by their corporate overlords.[14]

After *Uncanny Valley* was published, others were only too happy to pile on. Molly Young, a literary critic writing for *New York* magazine, agreed that Wiener's term—garbage language—was a well-chosen descriptor. As Young put it, "The hideous nature of these words—their facility to warp and impede communication—is also their purpose."[15] From this vantage point, it's only a short stroll to the language of Big Brother in George Orwell's *1984*: "War is peace. Freedom is slavery. Ignorance is strength." If Oceania existed, it would undoubtedly be awash in business buzzwords.

It may come as a surprise, therefore, to learn that corporate speak has its defenders. In a piece for *Slate*, Mark Morgioni responded to Young's jeremiad, arguing that so-called garbage language plays a useful corporate function. Morgioni's minority report is worth spending some time considering because it is a thoughtful refutation of the popular wisdom about jargon and the role that it plays in the workplace.

One advantage of corporate speak, Morgioni claims, is that its use saves time. As an example, he asks the reader to consider "stakeholder." This term clearly qualifies as a business buzzword: no one in a corporation below the level of executive would ever use it, except perhaps ironically. However, this one word efficiently refers to a specific set of people: those who would be most affected by a particular course of action and should therefore be taken into account and, if possible, involved in making a particular decision.

Morgioni offers a similar analysis of "parallel path"; once again, he claims that the phrase expresses a fairly complex idea in a simple way. His conclusion is that buzzwords offer a highly efficient means of communication, which is desirable when time is at a premium.[16]

As discussed earlier, jargon is often euphemistic, and such language can be used to conceal unpleasant truths. The flip side of this, however, is that it can also facilitate social interaction and face-saving.

As an example, Morgioni asks us to imagine a junior team member replying to a client's query with "Let me do a deep dive and circle back." Instead of having to admit to not knowing something, the phrase provides an escape hatch and assures the recipient that an answer will be forthcoming. In a similar way, "Let's put a pin in this" is a polite and socially acceptable way of terminating a fruitless discussion and allowing everybody to get back to work.

Finally, by speaking in this way, an individual displays familiarity with the conventions and customs of the tribe. By referring to "deliverables" instead of "products," she shows that she is part of the in-group and that she values her position within the organization. Viewed in this way, corporate speak can be thought of as a signaling device rather than as a medium for communication.

It would be easy to criticize Morgioni's support for "garbage language" by arguing that he cherry-picked more defensible examples of jargon to make his points. However, it might also be said that the buzzword haters choose to criticize the most egregious examples of jargon run amok. The truth probably falls somewhere between these two extremes. Corporate speak may well be reviled by many as unnecessary, self-important, or simply inane. At the same time, it may play important and perhaps less-well-understood communicative roles involving efficiency, group membership, and social desirability.

MISUNDERSTOOD AUTHORS

> *To be great is to be misunderstood.*
>
> —Ralph Waldo Emerson, *Self-Reliance* (1841)

> *People understand me so poorly that they don't even understand my complaint about them not understanding me.*
>
> —Søren Kierkegaard, *The Journal of Kierkegaard*

Writers complain about a great many things: the fickle nature of their muse, poor promotion by their publishers, and a world in which print is being supplanted by other media. But what may vex them most is not being understood by their readers.

To be fair, there are plenty of authors who don't make things easy for their audience. Many postmodern novelists deliberately forgo exposition, character development, or the story arcs that form plotlines familiar to readers of more conventional works. To cite two examples, Kurt Vonnegut made use of nonlinear time lines, while Vladimir Nabokov employed unreliable narrators. Although such techniques can achieve certain aesthetic effects, readers may find such narratives confusing or cognitively taxing. Other authors, like James Joyce (*Finnegans Wake*) or William Faulkner (*The Sound and the Fury*), have employed a stream-of-consciousness style that makes their books challenging to understand.

Authors can also try the patience of their readers with digressions. Extended passages that delve into minutiae, such as the intricacies of whaling (*Moby Dick*) or Parisian sewers (*Les Misérables*), are not especially reader friendly. And some authors, like Swiss psychologist Jean

Piaget, have a reputation for a lack of clarity that impedes the understanding of their ideas.[17]

On the other hand, the works of some authors are hard to fathom because they are grappling with complex or difficult ideas. German phenomenologists such as Kant, Hegel, and Husserl were attempting to communicate thorny philosophical concepts, and consequently their writing can be difficult to comprehend because of the subject matter.

To complicate matters further, the need to consider an author's intentions has been called into question. John Milton explicitly announced his objective at the beginning of *Paradise Lost* (1667): his epic poem is about justifying God's justice. And for generations, that is how his work was interpreted. However, later critics, such as William Blake and Percy Shelly, argued that the figure at the center of *Paradise Lost* isn't God but Satan.

Viewed from this perspective, Milton was mistaken about his stated purpose.[18] And some have gone farther than this, arguing that an author's intentions are irrelevant because the author is irrelevant. Roland Barthes, in his well-known 1967 essay, proclaimed the death of the author, arguing that it is the reader and not its creator who gives life to a piece of writing.[19]

Authorial intent and the reception of a work can be particularly problematic in the case of satire, which employs exaggeration, humor, and irony to criticize or to shame. Some works wear their satire on their sleeves: Heller's *Catch-22*, for example, contains outlandish elements that make it virtually impossible to miss its subversive, antiwar message. Other heavy-handed satires, such as Orwell's *Animal Farm*, slide into allegory: the novel's Mr. Jones is a transparent representation of Czar Nicholas II, Napoleon stands in for Stalin, Snowball is Trotsky, and so on.[20]

The intentions of other satirists, however, have been misunderstood. Consider the case of Daniel Defoe. Today, he is best remembered for having written *Robinson Crusoe*, one of the earliest English novels. At the beginning of his career, however, he was well known as a satirist and wrote about 50 such works during a 35-year period.[21] His most infamous work was the anonymous pamphlet *The Shortest Way with the Dissenters*.

The work was ostensibly an attack on those Protestants who disagreed with the Church of England. However, it made the case against them in a savage and exaggerated way and also cast the Tory ministry in a negative light. When Defoe's identity as its author was revealed, he was arrested and charged with seditious libel. Found guilty, Defoe was pilloried for three days and imprisoned for several months.

Modern scholarship is divided about Defoe's intentions in publishing the work as well as his subsequent defense of it. The pamphlet has been described as a "failed satire": as a counterfeit, it was too perfect. However, it was Defoe's reputation as a satirist that caused at least some of his contemporaries to read it as mockery of the Tory perspective.[22]

And the hall of mirrors regarding authorial intent doesn't end with Defoe. Some scholars have argued that Barthes's "death of the author" thesis was itself misunderstood and that his essay was a satire of the then-fashionable ideas of the New Critics.[23]

Parody is another genre in which confusion about the intentions of the author can occur. Parodies function as commentaries on another work and are typically intended to make fun of or to ridicule the original. Compared to the satirist, however, the parodist's job is somewhat easier. This is because parodists typically employ imitation of some sort to signal their intentions as clearly as possible. An unsuccessful parody is a work that fails to hew closely enough to the original to allow for such recognition.

Shakespeare's Sonnet 130, for example, has traditionally been read as a parody of love poetry made popular by Petrarch and then exploited by countless imitators writing about courtly love. Instead of following this convention and comparing his love to beautiful things like roses or stars, Shakespeare describes her hair as black wires growing out of her head and her eyes as "nothing like the sun."

Readers unfamiliar with anti-Petrarchan sonnets and this form of parody may mistake Shakespeare's intentions as mean spirited and come away with an understanding of the poem that is distinctly different from what was intended. Others, however, have argued that Shakespeare's description of his "Dark Lady" is a literal description of someone with dark hair and eyes who was perhaps of Mediterranean or Jewish descent.[24]

As these examples illustrate, it is all too easy for writers to miss their marks when attempting to communicate their ideas. At the same time, readers bring their own preconceptions about an author, a genre, or a narrative structure to the act of reading. Perhaps Emerson should have said, "To *write* is to be misunderstood."

LOST IN TRANSLATION

Although the focus of this book is on miscommunication between individuals sharing the same language and culture, there are also significant crosslinguistic issues that can derail communication or create misunderstandings. An important factor is the act of translation itself, which raises a host of thorny lexical, semantic, and cultural issues.[25] There are many infamous examples of mistranslation, and three of these are described below to illustrate the range of such errors—from the trivial to the consequential.

The Canals of Mars

During the Great Opposition of 1877, when Mars and the Earth were separated by only 35 million miles, a number of discoveries were made about the Red Planet. Using a telescope at the U.S. Naval Observatory, Asaph Hall was the first to detect Mars's two satellites. And in Milan, Giovanni Schiaparelli took advantage of the opposition to carefully study the planet's surface with the Brera Observatory's 22-centimeter instrument.[26]

Schiaparelli's observations were the basis for his map of Mars, which contained a number of dark areas that he thought might be oceans. The astronomer gave them names, like Syrtis Major, that are still in use today.[27] However, Schiaparelli also reported seeing a number of straight dark lines running across the planet's surface and referred to them as "canali," the Italian word for "channels." This term, however, was translated into English as "canals." And whereas the term "channel" is agnostic about the provenance of such features, "canal" implies some sort of intelligent design.

A few years later, the American amateur astronomer Percival Lowell advocated for the notion that these canals were artificial structures, built by the inhabitants of Mars to transport life-giving water from the planet's polar ice caps to the thirsty denizens of the equatorial regions. The canals were visible from Earth, he asserted, because verdant strips of vegetation had sprung up alongside these structures. Such notions were entertaining to a credulous public, and a best-selling book that Lowell wrote inspired H. G. Wells to pen his science fiction classic *The War of the Worlds.*

Lowell attempted to obtain photographic evidence for the canals during another planetary opposition in 1907, but the small and grainy images he obtained did little to sway his doubters.[28] Later orbital missions and landers conclusively proved that such features were an optical illusion: there are no canals on Mars, and the dark spots observed by Schiaparelli were exposed basaltic rock instead of oceans.

The Horns of Moses

The basilica of Rome's San Pietro in Vincoli is a popular destination for tourists visiting the Eternal City. The reliquary under the main altar is said to contain the chains of Saint Peter, given to Pope Leo I by Empress Eudoxia in the fifth century. The church also houses the tomb of Pope Julius II. Due to a lack of funds when it was constructed, its original grand design was scaled back to only a handful of statues. One of these, however, is Michelangelo's sculpture of Moses, an unrivaled work of art by the Renaissance master.

Michelangelo was justifiably proud of his creation, which is exceptionally lifelike—except for the fact that Moses is sporting a pair of horns on his head. Why did he include such an incongruous detail? Michelangelo's reasoning was simple: Moses is described in the Bible as possessing horns—according to one translation, at any rate.

The passage in question is from Exodus chapter 34, verses 29–30: "When Moses came down from Mount Sinai with the two tablets of the covenant law in his hands, he was not aware that his face was radiant because he had spoken with the Lord. When Aaron and all the Israelites saw Moses, his face was radiant, and they were afraid to come near him."

This is from the New International Version, produced by scholars working from the earliest Hebrew sources for the Old Testament.

However, during the lifetime of Michelangelo, the Vulgate version of the Bible, translated into Latin by Saint Jerome more than 1,000 years earlier, was the preferred version. And Jerome translated the Hebrew קרן as "keren" or "horned" ("cornuta" in Latin). However, it could also be read as "karan," which is written with the same Hebrew letters but pronounced with different vowels. And "karan" means "radiant" or "beaming."[29]

Although Michelangelo was not the only artist to depict Jews with horns, such portrayals and their devilish connotations helped to fuel anti-Semitic sentiments for centuries.

The Threats of Khrushchev

On November 18, 1956, Soviet leader Nikita Khrushchev spoke to a group of NATO envoys during a reception held at the Polish embassy in Moscow. The event occurred at a particularly strained time during the Cold War, coming on the heels of the Hungarian Revolution and the Suez Crisis. East–West relations were at a low ebb, and tensions were high.

During the course of some unscripted remarks about capitalist nations, the first secretary asserted, "Whether you like it or not, history is on our side," and concluded, "My vas pokhoronim!" This phrase, widely reported as "We will bury you," was interpreted by many in the West as a naked threat to use nuclear weapons.

Years later, Viktor Sukhodrev, who had served as Khrushchev's interpreter, spoke about the event in an interview with the RT network. Although he asserted that "We will bury you" was a strict interpretation of his employer's words, he also admitted that Khrushchev often enlivened his impromptu speaking with Ukrainian proverbs with which the translator was unfamiliar.[30] And it is possible that the Soviet leader was echoing a Russian aphorism, which translates, more or less, as "We'll be here even when you're dead and gone."[31]

In other words, Khrushchev's point may have been that communism would outlast capitalism and not that his goal was to destroy the West.

Another possibility is that he meant the phrase metaphorically—that ultimately, communism would triumph over capitalism as an economic system.[32]

Whatever Khrushchev intended by his remark, Western politicians seized on it as a threat, interpreting it literally and repeating it often. For the rest of the Cold War—through the Bay of Pigs invasion, the Cuban missile crisis, the building of the Berlin Wall, and beyond—the remark hung in the air, poisoning East–West relations. It would take 30 years, until Mikhail Gorbachev and the glasnost era of the mid-1980s, for Khrushchev's assertion to fade into history.

ECHO CHAMBERS AND CUCUMBER TIME

In the middle of June 2019, the news media spent a few cycles trumpeting the idea that cell phone use was turning millennials into monsters—specifically, by causing them to grow horns.

The research that journalists were reporting had been published in February 2018 in *Scientific Reports*, a peer-reviewed online journal. Two Australian health scientists examined more than 1,000 x-rays of skulls and reported that about a third showed "prominent exostoses"—bony outgrowths—on a specific area near the base of the cranium.

The researchers speculated that screen-based activities might be the reason for the growths, with the cause being increased mechanical stress as individuals crane their heads downward to stare at phones or tablets. These exostoses, then, could be an adaptive response to changes in posture. The researchers found that the growths were more commonly seen among those aged 18 to 30 and that they were more than five times more common in men.[33]

The study was given a new lease on life some 16 months later when the BBC posted an online article titled "How Modern Life Is Transforming the Human Skeleton." That article referred to the exostoses as "spiky growths."[34] But it was a subsequent *Washington Post* story on this research that referred to them as hooks or horn-like features, with the title of the piece asserting that "'Horns' Are Growing on Young People's Skulls. Phone Use Is to Blame, Research Suggests."[35]

And the scare quotes are entirely missing from the *New York Post*'s headline: "Young People Are Growing Horns from Cellphone Use."[36] Other news outlets, such as NBC and *Newsweek*, ran similar stories.

It's worth noting, however, that the term "horn" did not appear in the original research paper. And even though the word came up in subsequent interviews with the authors, it was stressed that this was in reference to the *shape* of the bone spurs[37] and not their composition: horns, after all, are made of a substance called keratin, not bone.

Perhaps inevitably, there followed a wave of news reports debunking the claim that younger adults were growing horns on their heads.[38,39] The original paper was criticized on a number of grounds, such as the fact that the researchers didn't measure cell phone use. In addition, the lead author was identified as someone who stood to benefit financially from attention paid to this topic: he was a purveyor of corrective devices and pillows marketed under the moniker "Dr. Posture."[40]

In the case of horns and cell phones, a number of media outlets acted quickly to discredit the more lurid claims that were being made. In other cases, however, sensational reporting has deeply implanted dubious findings into the public's consciousness and even affected public policy.

For example, a report in the journal *Nature* in 1993 demonstrated that college students briefly performed better at a spatial reasoning task after listening to a Mozart sonata.[41] Throughout time, this morphed into the "Mozart effect": a belief that exposing infants and children to the Austrian composer's compositions would confer lasting cognitive benefits. Such claims were advanced primarily by author Don Campbell, but in this case, the pushback from journalists was muted.

By 1998, Governor Zell Miller was including funds in the Georgia state budget to provide the parents of newborns with a CD of classical music.[42] It was left largely to research psychologists to push back against the Mozart effect by showing that the initial finding consistently failed to replicate.[43]

To be clear, frivolous or inaccurate reporting by the news media is not something new under the sun, although it used to be more seasonal. Long before contemporary concerns about fake news, inane or irresponsible reportage could often be found in print, particularly during the so-called silly, or slow news, season. This corresponds roughly to the dog days of summer, particularly July and August, when legislative bodies are not in session and other news makers are on holiday. (In many countries, this period is referred to as "cucumber time."[44])

Concerns about the miscommunication of scientific results to the public has a long history as well. Unease regarding how journalists report research on topics like health studies and climate change began to appear in scientific journals as early as the 1990s.[45,46] One paper explicitly compared the media's distortions of a committee's findings about climate change to Chinese Whispers—the Telephone game described in chapter 7.[47]

Studies of why such miscommunications occur have highlighted a variety of causes. Among these are the perverse incentives and rewards of sensationalism for both researchers and journalists.[48] The reputations of scientists benefit from increased visibility, and publications that make sensational claims benefit economically.

The shifting media landscape and its new business model is also a factor. Instead of attracting newsstand customers and subscribers, many publications chase the eyeballs of online readers to expose them to advertising. Even well-respected media outlets have been known to publish stories that are essentially clickbait: reports with provocative headlines but little in the way of responsible journalistic content. The end result, however, may be a weakening of the public's faith in institutions like journalism as well as science itself.

COME HERE OFTEN?

> *Inexperienced males can occasionally be seen to pick the most physically appealing females; the more experienced males seem to know better and try to join the females who look to them.*
>
> —Mark S. Carey, *Nonverbal Openings to Conversation*, 1974

Clearly, some communicative acts are more consequential and parlous than others. It can be difficult, for example, to find the exact words to formulate a sincere apology or, for that matter, a veiled threat. And when we have to impose on other people, we must also choose our words carefully.

Another highly fraught discourse domain is the communication of romantic interest, followed by gauging the reaction to such overtures from clues provided by the recipient. The frisson of excitement that one feels when approaching a potential partner can easily dissolve into misunderstanding, confusion, and embarrassment. Affairs of the heart, unfortunately, seem to be tailor-made for miscommunication.

One reason that flirtation frequently leads to misunderstanding is that, by its very nature, it is an ambiguous form of communication.[49] An explicit come-on is likely to be perceived negatively, but one that is too subtle may not even be perceived as an expression of interest. At the same time, receptivity that is too quick or emphatic can make a person appear somewhat desperate or possibly indiscriminate. These are qualities that tend to be unappealing. It may be better to be perceived as hard to get, even at the risk of coming across as appearing uninterested. But enigmatic conversation starters, as well as the vague responses they elicit, are clearly a strike against clear and effective communication.

Evolutionary psychologists have argued that it is in the best interests of men to approach as many potential partners as they can. Flirtation, for them, can be thought of as a numbers game—and a pickup line that falls flat with one potential target might well work on another. Diligent efforts should lead to success sooner or later.

According to evolutionary theory, the issues for women are quite different, and they must be choosier. After all, if a dalliance leads to a sexual encounter and pregnancy, a great deal of time and effort must be expended on child rearing since that role has traditionally fallen to mothers.[50] Therefore, the potential stakes in responding to flirtatious behavior are higher, and women need to be more discriminating.

Interest in flirtatious behavior has driven researchers into bars—not to drink, necessarily, but to study such doings in the wild. The effectiveness of various pickup lines, for example, has been assessed in such venues. Researchers have found that opening gambits classified as cute or flippant ("You remind me of someone I used to date" or "Bet I can outdrink you") are received more negatively by female patrons than those classified as direct ("I feel a little embarrassed about this, but I'd like to meet you") or as innocuous ("Hi" or "What do you think of the band?").[51]

As the quote by Mark Carey at the beginning of this section suggests, the signals of receptivity in such venues are often nonverbal in nature.[52] Not surprisingly, perhaps, science has demonstrated that multiple bouts of eye contact in a bar led to higher rates of "approach behavior" than single episodes. The researchers interpreted this as suggesting that "males may need certain encouragement before approaching a female stranger."[53]

Other fieldwork has investigated whether the *length* of a returned glance, accompanied by a smile, predicts the approach of a potential suitor (spoiler alert: it does).[54] And a singles bar in an Ontario college town provided researchers with real-world examples of SOABs (sexually overt approach behaviors), such as the touching of the breasts or buttocks or grinding with one's dance partner.[55]

As in other domains, people have fairly detailed scripts they make use of to interpret the behavior of others. Sexual scripts are no exception to this, and they include expectations that provide a framework for flirtation and romantic encounters.

Certain behaviors, such as physical contact, are seen as signals that make a sexual encounter more likely. And both men and women agree that the likelihood of sex increases when the locale moves from a public to a private setting. Other behaviors, such as expressions of concern on the part of the woman ("I think this is moving too fast"), are perceived by both genders as decreasing the likelihood of a sexual encounter.[56]

The sexual scripts of men and women may largely overlap, but even minor differences can lead to miscommunication and misunderstandings. When it comes to the motivations for flirting, for example, men view it as being more sexual, whereas women are more likely to perceive it as a harmless way of having fun.[57] Interestingly, however, when men are perceived as flirting for fun, their attractiveness to women decreases significantly.[58]

Another source of misunderstanding in this domain is that friendly behavior can be misperceived as flirtatious behavior and vice versa. This can be especially problematic in the workplace, where most coworkers try to develop positive relations with their fellow employees. Many signals of simple camaraderie, such as friendly chitchat or remembering a coworker's birthday, may be completely innocent. However, they can easily be misinterpreted by the recipient as something more consequential, such as a signal of romantic interest.

Perhaps most significantly, miscommunication can lead to misunderstandings regarding sexual consent. The #MeToo movement has shone a spotlight on how those in dominant or powerful positions can abuse their privilege. A shift to explicit affirmative sexual consent has the potential of reducing unwanted sexual contact by decreasing the ambiguity that is created by passive consent.[59]

10

CONTEXTS, PART 2

In the final chapter of this book, we return to the issue of egocentrism to help explain why emails can be hard to interpret correctly. And even phenomena that may seem far removed from communication, like road rage, can be understood as arising from miscommunication. Similar issues of misunderstanding affect the credibility of witnesses in the courtroom and have even caused slaughter on the world's battlefields. We conclude by considering the specialized language used in aviation to prevent accidents and a disaster from 1977 that tragically demonstrates the fragility of communication when multiple safety systems fail.

EGOCENTRISM AND EMAIL

An MBA student said about emails from her boss: "I can never tell how my manager feels. When organizing a meeting I got a sarcastic reply ('this had better be good') that I took to heart." Only much later did she find out the manager intended the comment to be funny.

—Kristin Byron (2008)[1]

One survey respondent said, "I wrote a question to (my boss) one day; she thought I was being insubordinate by the tone. I almost lost my job!"

—Sarah Schafer (2000)[2]

What is it about electronic messages that make them open to such mis-interpretation? To some degree, this shouldn't be surprising by now. As we saw in the discussion of egocentrism in chapter 2, people asked to tap out the rhythm of a song mistakenly believe that others will be able to identify it. And in that chapter, we also saw how the curse of knowledge blinds us to detecting errors, such as Moses instead of Noah bringing animals onto the ark. These types of cognitive biases create issues for communication in general, but they help to create a perfect storm when communicating via text or email.

In addition to these problems, email lacks many of the cues that can disambiguate face-to-face conversations. We can't, for example, rely on facial, gestural, or vocal cues to help us get our meaning across. We can't see a puzzled frown that would serve as a prompt to clarify our intent. And we can't perceive a nervous reaction that would lead us to offer words of encouragement. Email is an impoverished medium that doesn't allow for such real-time feedback.

A number of studies have documented that we are overconfident in communicating our intentions via email. Justin Kruger and his collabo-rators asked college students to write both serious and sarcastic emails and to indicate whether someone else would be able to correctly deter-mine their intent. The college students also classified the emails of other participants in the study. When writing the emails, the students thought their serious or nonserious intentions would be clear to another person 97 percent of the time. When evaluated by others, however, the accuracy rate was 84 percent—still fairly high but significantly lower than what the students expected.[3]

The participants in this study were strangers to one another. Would friends do any better? After all, our friends are familiar with how we express ourselves. We can also rely on the common ground that we share with them. So it seems as if friends should have an advantage in tasks like these. In a follow-up to Kruger's study, Monica Riordan and Lauren Trichtinger had participants write emails that expressed particular emo-tions to both friends and strangers. Even though the participants thought their friends could figure out their affective states more accurately than strangers, there was no difference in performance between the two groups.[4]

Another study was designed to assess the reliability of classifying emotions in a group of email messages. In general, the participants were "unreliable and inconsistent" in detecting the emotions in the messages they evaluated.[5]

Clearly, egocentrism and the curse of knowledge can help to explain these results. But other factors may be at work as well. Kristin Byron has suggested that emails containing positive affect are perceived as less emotional than their authors intend them to be, a phenomenon referred to as the neutrality effect. It is thought to occur because, compared to face-to-face or even telephone conversations, email is less physiologically arousing.[6] Viewed from this perspective, positively valanced words on a screen simply lack the emotional resonance and punch they possess when delivered by a flesh-and-blood person.

In addition, Byron suggests that emails may be perceived more *negatively* than intended by their senders. This negativity effect may be due, in part, to the relative brevity of many messages, which can make them come across as brusque or curt. Negativity effects can lead to a number of unintended consequences, such as conflict escalation or anxiety about one's performance, as seen in the two examples at the beginning of this section. As Byron put it in the title of her article, we may be asking electronic messages to carry a communicative burden that is too heavy for this medium to handle.[7]

And if we zero in on a particular form of communication, we see once again how email is less effective than face-to-face interaction. We often use language to persuade others—for example, to change their opinion or to ask for a favor. So should we make requests of others in person or by email?

Imagine that you approached 10 strangers at random and asked them to fill out a questionnaire. Out of the 10, how many do you think would do it? If you guessed "about half," your intuitions are in line with the 49 research participants who were asked this question—their average estimate was 5.08. However, when each of these college students went out on campus and asked 10 people to fill out the form, the average compliance rate was 7.15, or 40 percent higher than they expected.

The same participants were asked about compliance if they instead sent an email request to 10 people. The students estimated that a little more than half (5.53) of 10 people would fill out the form. But when they sent requests to 10 email addresses chosen from a university directory, the compliance rate was only .21, or 26 times *less* than they expected![8]

In hindsight, the results make sense since it's hard to ignore an in-person request and relatively easy to ignore an email. But the discrepancy between what the students expected and then observed in the email condition is striking. The authors of the study posit that their research participants "failed to appreciate the suspicion, and the resulting lack of empathy, with which targets view email requests from strangers."[9]

Taken as a whole, the results from these studies portray electronic communication in a relatively poor light. However, many of the problems revealed by researchers simply reflect the biases that people bring to almost any communicative situation.

RAGE ON THE ROAD

In December 1997, Cheryl Kyle and her husband Robert were driving home after enjoying a Christmas dinner with relatives in The Dalles, Oregon. On Interstate 205 in Portland, Cheryl started to make a lane change and accidently cut off another motorist. Although she worked as a school bus driver and was always careful to check behind her, the small brown-and-tan Toyota had been in her blind spot. She pulled the couple's pickup back into her original lane while the driver of the Toyota sped up alongside them.

Cheryl and her husband mouthed the word "sorry" and held up their hands in a way that suggested regret. Their words and gestures, however, failed to mollify the other driver, a middle-aged man with curly hair. His response to their apology was to roll down his window and shoot at them. A bullet from a 9mm handgun hit Robert's arm just above the elbow. In recounting their story to the police, Cheryl recalled that after firing at them, the man had simply driven off as if nothing had happened.[10,11]

Road rage is common enough that a 2003 telephone survey—of nearly 1,400 notably polite Ontarians—found that nearly half of them reported having been shouted at, cursed at, or on the receiving end of rude gestures during the previous year. Almost a third of these respondents reported having engaged in such behavior as well. Seven percent had been threatened with harm to their vehicles or themselves, although only 2 percent admitted to making such threats.[12] Perhaps not surprisingly, those who express anger in an aggressive manner tend to be male and younger.[13]

Although the concept of road rage has probably existed for as long as people have traveled in vehicles, the term itself is fairly recent, dating from 1988.[14] One reason it occurs is that drivers on the open road have a fairly limited behavioral repertoire with which to communicate. They can mouth or mime expressions of regret, like Cheryl and Robert Kyle, but such gestures lack nuance and may be perceived as perfunctory instead of sincere. And aggrieved recipients, in vehicles that confer a certain degree of anonymity, can transgress against those who they believe have wronged them with little fear of reprisal.

All cars and trucks are equipped with rudimentary communication systems, such as turn signals and horns, while others, such as headlights, may be pressed into service as well. Although turn signals provide unambiguous signs of intent, the same cannot be said for horns. Most drivers attempt to make the briefest of taps on their steering wheels to alert another motorist that, for example, a stoplight has changed from red to green. Anything more than a fleeting honk is liable to be perceived as hostile, which can engender anger in return.

The use of headlights to communicate is even more problematic since motorists may flash them for a variety of reasons. This is done, for example, to request oncoming drivers to switch from using their high to low beams at night. Truck drivers may also flash their lights to indicate that another vehicle can now merge safely in front of them. At one time, it was common to flash one's lights at other motorists to indicate that they should turn on their own headlights. However, a persistent urban legend—that gangs use this behavior to identify victims for initiating new members[15]—has largely put an end to such signaling.

Drivers can also flash their headlights to warn others about a police speed trap. In the United States, laws about the legality of such signaling vary from state to state. In some jurisdictions, flashing one's lights for this reason can result in a ticket for interfering with a police investigation. However, many courts have ruled that headlight flashing constitutes protected free speech under the First Amendment.[16]

In addition, many drivers flash their lights to indicate they want to overtake a slower motorist. This can be perceived as a hostile act, particularly when accompanied by tailgating of the slower vehicle. Drivers who are aggrieved by such antics may slow down and tap their brakes to briefly illuminate their brake lights as a warning to the tailgater.

For good or ill, there is a universally recognized finger gesture for communicating anger. There is, however, no corresponding gesture for signaling "I'm sorry" or "My fault." A brief wave, for example, might not be seen or may be perceived as insufficiently contrite. And as we saw with the Kyles, a supplicating, palms-raised gesture may not be sufficient to mollify someone who is prone to road rage. There have been proposals to create such a gesture, such as a light tap of the palm to one's forehead.[17]

Once again, we see the brittleness of communication when the methods we have for expressing ourselves are impoverished or constrained. As soon as we climb behind the wheel, we have a strike against our ability to communicate. A number of patent applications have been filed that aim to solve this problem, such as an "apology stick"—a remote controlled flashing sign—or a hand-shaped structure, mounted on the back of a vehicle, that can be made to move from side to side to simulate a wave. Tom and Ray Magliozzi, of NPR's *Car Talk* fame, weighed in on this issue in 2001. Their take was that drivers are already too distracted and that adding another communication channel would ultimately be counterproductive.[18]

The most intriguing proposal to solve this problem may be the development and adoption of a special "I'm sorry" horn.[19] This would expand the spectrum of audible cues from the always-negative honking to more beneficent acts of thanking or contrition. It might reduce miscommunication on the road—or simply provide another way to signal sarcasm by those who are not, in fact, sorry.

DISORDER IN THE COURT

The legal profession does its best to impose order, clarity, and fairness onto the unruly world of human affairs. And the application of the law can involve looking into the hearts and minds of individuals to divine murky abstractions like motivation and intent. Consider, for example, the distinction between murder and manslaughter.

Murder is an act committed with "malice aforethought," while voluntary manslaughter involves some sort of provocation. This is a distinction that requires juries to grapple with nebulous concepts like "moral blame" and the "heat of passion."[20] The verdicts reached in cases involving homicide have significant consequences as well. Because the punishment for manslaughter is less than that for murder, a prison sentence for murder can be significantly longer.

To serve justice, legal practitioners have to grapple with and guard against misunderstanding and miscommunication in a variety of contexts in the courtroom. To begin with, judges and juries must consider the evidence provided by the opposing legal teams during a trial. As we saw in chapter 8, the admissibility of hearsay evidence in the courtroom is one way that the legal profession attempts to separate opinion from fact. Distinctions are also made between points that all parties agree to (stipulations) and those that are in dispute (allegations).

In addition, prosecuting attorneys must often obtain testimony from individuals who are trying not to implicate a friend, a loved one, or themselves as having been involved in criminal activity. So-called hostile witnesses may deliberately misunderstand an attorney's questions or interpret them more narrowly than intended.[21] They may be generally uncooperative or refuse to respond at all by invoking their right not to incriminate themselves.

Despite the best efforts of everyone involved, communication problems can still arise. Judges, for example, have the responsibility of applying a sometimes ambiguous legal code to the behavior and misbehavior of those who stand before them in their courts.

Confusion can also arise from the instructions provided to juries by judges. It may be necessary, for example, for the judge to define legal concepts in ways that are different or more precise than how such terms are used more generally. A good example is the concept of "proof beyond a reasonable doubt": this has been a contentious issue in jurisprudence since at least the eighteenth century,[22] and it has been claimed that such instruction can introduce a bias toward conviction.[23]

The judicial system has erected a number of safeguards that are designed to prevent miscommunication and misunderstanding with regard to the trial itself. One of these is the presence of a court reporter or stenographer who creates a real-time transcript of the words spoken by all parties and who stands ready to review this record if necessary. In a perfect world, this would be a flawless record of the court's proceedings. But because communication is complex, the transcription of a legal proceeding may be imperfect in a variety of ways.

This issue came to the fore in 2013 during the court case of George Zimmerman, who was on trial for the death of Trayvon Martin. A crucial witness in the proceedings was Rachel Jeantel, a close friend of Trayvon's who was speaking with him by phone when the tragic events began to unfold. The young woman, still grieving, provided soft-spoken answers in African American English (AAE).

AAE is a dialect that differs in specific ways from other varieties of English, such as in how tense, aspect, and mood are indicated. For example, AAE allows the deletion of the verb "to be" in the present tense, so "He workin'" means "He is working," whereas "He be workin'" is equivalent to "He usually works" or "He is often working."

In addition, there are sound differences, such as pronouncing "three of those" more like "tree of doze" and "bath" as "baf."[24] During Jeantel's testimony, the court reporter and members of the jury, who were unfamiliar with AAE, often struggled to understand her.[25]

The Trayvon Martin trial inspired language researchers to determine whether the accuracy of court reporters might be affected by the transcription of vernacular forms of English. They asked 27 court reporters who were working in Philadelphia courtrooms to transcribe and paraphrase speech that included features of AAE.

To be certified to work in the courts, these reporters must achieve a 95 percent level of accuracy at rates of speech of at least 225 words per minute.[26] (For purposes of comparison, this is a rate three times faster than required for professional typists.) Court reporters achieve such impressive levels of speed and accuracy by using machines with special keyboards. By using combinations of keys, they can quickly record syllables, words, and whole phrases.[27]

The court reporters in the study heard 83 recorded utterances of AAE that were drawn from courtroom testimony. After each one, they were asked to provide a transcription and to paraphrase what the speaker meant. Their average transcription accuracy was only 60 percent—far below the 95 percent level required for professional certification. Even the best reporter achieved only 77 percent accuracy, while the worst was at 18 percent.

Paraphrase performance was even worse, with an average accuracy of only 33 percent. Many of the reporters found this task to be extremely challenging and expressed frustration in their attempts to arrive at the speaker's intended meaning.

One of the study participants, for example, told the researchers, "The tenses drive me crazy! *He be workin'*: what does that mean?! He *is* working? He works? He does work? That drives me *nuts!*"[28] And the Black court reporters, who were in many cases unfamiliar with AAE, performed no better than their white counterparts.

The researchers concluded that the reporters' training simply hadn't prepared them for the nonstandard dialects of English that they encounter in the courtroom. And since these transcriptions become the official record of the trial, any mistakes are consequential: they introduce errors that may disadvantage nonstandard speakers in a variety of ways, such as in a retrial. Justice may be blind—but it also doesn't hear all that well.

MILITARY MISADVENTURE

> *"Forward, the Light Brigade!"*
> *Was there a man dismay'd?*
> *Not tho' the soldier knew*
> *Someone had blunder'd.*
>
> —Alfred Tennyson, "The Charge of
> the Light Brigade" (1854)

The events described in Tennyson's well-known poem occurred during the Battle of Balaclava in the second year of the Crimean War. Toward the end of that battle, the British commander, Lord Raglan, ordered an attack to be made by a brigade of light cavalry.

Raglan's intention was to keep the opposing Russian forces from making off with the naval guns they had captured earlier, when they occupied positions previously held by allied Turkish forces. The British cavalrymen, armed only with sabers and lances, were well trained and equipped for such a mission.

Instead of chasing off the Russian infantry, however, the Light Brigade made a frontal assault on a different part of the field, encountering heavy artillery fire from multiple directions. Despite the hailstorm of projectiles tearing through the ranks, the advance continued, and a few dozen horsemen eventually reached the Russian guns. By this point, however, the brigade's ranks were so depleted that the survivors had no choice but to withdraw.

Exactly why this tragedy occurred is still a matter of some dispute. Lord Raglan's orders were delivered by Captain Louis Nolan to the cavalry commander, the Third Earl of Lucan. The orders read, in part, "Try to prevent the enemy carrying away the guns." The artillery in question could not be seen from Lucan's position, and he angrily asked Nolan to specify which guns were meant.

Nolan, for his part, was exasperated by Lucan's slow reading of the orders and his apparent confusion. Instead of making clear the location of the captured naval guns, Nolan made a sweeping gesture, apparently toward a different Russian position about a mile away. He is reported to have said, "There is your enemy! There are your guns!"[29]

Nolan appeared to be pointing to the end of a dale that would later be made famous by Tennyson as the "valley of Death." The Russian forces had more than 50 artillery pieces dug in on both sides and at the end of this valley, with unobstructed views of any approaching forces. Nevertheless, Lucan relayed the order to the Light Brigade commander, Lord Cardigan, who questioned the wisdom of such an attack. Lucan replied, "We have no choice but to obey."

The British assault on this position and the attendant butchery lasted only about seven minutes. Of the 670 cavalrymen who made the attack, more than 300 were killed, wounded, or captured. Nolan himself charged to the front of the brigade and was killed almost instantly.

He may have recognized his earlier miscommunication and was trying to correct it, but if that was the case, he did not survive long enough to do so. The Russians soldiers, for their part, were so astonished by the suicidal attack that they concluded the British forces must have been drunk.[30]

Since the beginning of organized warfare, military commanders have struggled to communicate effectively with their troops. The hierarchical organization of armies and navies greatly increases the chances for Telephone game effects to occur, and the Charge of the Light Brigade is a tragic example of this.

As recently as the mid-nineteenth century, a commonly employed solution to communication difficulties was to have generals personally lead their soldiers into battle. This certainly afforded commanders the opportunity to see what was happening and to perhaps inspire their forces, but such advantages came at a high cost.

During the American Civil War, for example, about 75 Confederate generals were killed or mortally wounded. This was out of a total of 425 who held that rank, or more than one in six.[31] The loss of that many senior commanders was unsustainable and undoubtedly contributed to the South's military disadvantages. In addition, advances in technology and communications since that time have made it unnecessary for commanders to place themselves directly in harm's way.

However, not all casualties are caused by the enemy. The fog of war created by battlefield conditions has many consequences, including imperfect communication between allied forces. The phenomenon of so-called friendly fire has probably existed for as long as warfare itself, and miscommunication can often be found at the heart of such episodes.

For example, one of the Confederacy's most able commanders, Thomas "Stonewall" Jackson, was cut down by his own troops during the battle of Chancellorsville. Returning to quarters at dusk, he and his staff were challenged by soldiers of the 18th North Carolina Infantry, who fired on them even as they frantically tried to identify themselves. Suspecting that it was a trick, the soldiers fired again, and Jackson was hit by three bullets. His left arm was amputated, and he died from complications of pneumonia eight days later.[32]

The phenomenon of friendly fire shows no signs of becoming less frequent. For example, a lack of communication between the army and navy during the Falklands War in 1982 led to the HMS *Cardiff* shooting down a British Gazelle helicopter, resulting in four deaths.

During the Persian Gulf War in 1991, American forces sustained 615 casualties—a remarkably low number. However, 23 percent of those casualties (35 killed, 72 wounded) were the result of friendly fire. The same phenomenon accounted for more than three-quarters of the combat vehicles that were destroyed during the conflict.[33]

The death of the popular football player Pat Tillman in Afghanistan in 2004 led to renewed concerns about the causes of friendly fire incidents. However, episodes of miscommunication involving drone strikes in that country suggest that the problem persists.[34]

THE LANGUAGE OF THE AIR

Although a fear of flying is widespread, modern civil aviation is incredibly safe. Between 2012 and 2017, there were 228 million flights worldwide, but only 419 accidents. Moreover, only 15 percent of those accidents resulted in fatalities. And in 2017, when there were nearly 42 million flights, 19 people died in plane crashes.[35] To put it a different way, your odds of being a passenger on a plane crash with fatalities are less than one in 3.8 million. You are far, far more likely to succumb from bee stings or being struck by lightning.

The accidents and near accidents that do occur result from of a variety of causes, such as severe weather or equipment failure or malfunction. However, about 80 percent of such incidents are thought to be the result of pilot or air traffic control errors.[36] And although such incidents are typically the result of many factors, miscommunication and misunderstandings often play a role.

Although many airlines operate within only one country, a significant number of daily flights are international. As a result, the language spoken in the destination nation is often different than that used at the point of departure. And just as a common pidgin language developed to aid commerce in the Levant hundreds of years ago, English has come to function as the lingua franca of the air. English was spoken by some of aviation's pioneer pilots and manufacturers, but it is also widely spoken throughout the world due to its use in former British colonies.

The use of English in aviation was codified by the International Civil Aviation Organization (ICAO), an agency of the United Nations. Headquartered in Montréal, it oversees standards such as airport and airline codes as well as aircraft registration and type designators. In 1951, it recommended the universal adoption of English for plane-to-ground communications.

This recommendation was revised in 2003, when the ICAO announced that, within five years, aviation professionals would need to demonstrate proficiency in English. Specifically, the ICAO requires pilots and controllers to demonstrate English-language competence to at least the fourth level of a six-level proficiency scale. Periodic reassessments are also required for those whose proficiency is below level six. Unfortunately, there is widespread concern about the assessment procedures that are employed as well as cheating on proficiency exams and certificates being issued illegally by corrupt officials.[37,38]

The language of the air isn't the same as the English spoken by civilians, however. Aviation English is an example of what is called English for Specific Purposes (ESP), varieties of which are taught worldwide. In ESP programs, individuals in fields such as medicine, business, or tourism receive instruction in a relatively narrow vocabulary to allow them to function effectively in multilingual settings.

Aviation English consists of about 300 words, and most of these are used in highly specific contexts. "Pan-pan," for example, is uttered when a situation is serious but not life threatening ("mayday" is reserved for when people are in jeopardy). Such phrases must be repeated three times at the beginning of a radio call.

Given the need for absolute clarity, Aviation English differs in several other ways from standard English. Because letters can easily be misheard, particularly in a noisy cockpit or ground control setting, they are indicated by specific words that start with that letter; for example, "f" and "s," which are often confused, are made distinct by reference to "foxtrot" and "sierra."

Many languages lack the sound at the beginning of "three" and "thousand," and nonnative speakers can find it challenging to produce. As a result, these numbers are pronounced as "tree" and "tausand" instead. Numeric specifications are spelled out completely: 1,300 would be communicated as "one tausand tree hundred" as opposed to "thirteen hundred" or "one-tree-oh-oh."

In Aviation English, clarity is more important than conciseness. Important elements of clearances or instructions must be read back to ensure they were heard correctly. Negative constructions are avoided. Errors must be followed by the word "correction" and then the right information. In general, the communication between the pilots and ground controllers follows a predictable script since problematic communication is more likely to occur when either party deviates from established protocols.[39]

As a result, even if someone is a native English speaker, that does not mean that she is, by definition, proficient in Aviation English. And since it is only a subset of standard English, native speakers may spontaneously make use of nonstandard terminology that nonnative pilots or controllers won't understand. In emergency situations, people tend to speak rapidly, and this can also be problematic for those who are not proficient in English. Finally, nonnative speakers who *are* proficient in Aviation English may speak it with a pronounced or unfamiliar accent, and this can also hinder or impair communication.[40]

Some of the culprits for problematic aviation communication may be idiosyncratic to Aviation English and the complexities of flying and landing aircraft. Others, such as ambiguity and homophony,[41] occur in everyday discourse, as we have seen in previous chapters. Laboratory studies, in which pilots and controllers converse during simulated flights, have revealed other issues.

Not surprisingly, native speakers of English and those with higher levels of flight qualification tend to make fewer communication errors. The performance of nonnative speakers, while lower overall, is also affected when transmissions such as read-backs or reports lack pauses.[42] Pauses can function to highlight important information, and they also provide less proficient speakers with time to process what they are hearing.

Some aspects of Aviation English might appear to be common sense. Others might seem like needless verbosity. But all of it is meant to reduce misunderstandings between flight crews and ground controllers since errors can have devastating consequences. The final section of this chapter describes a tragic example of this.

TRAGEDY AT TENERIFE

On March 27, 1977, the ground controllers at Los Rodeos Airport were having an extremely busy afternoon. The regional airport, located in the Spanish Canary Islands off the coast of Africa, was being inundated with incoming planes.

The heavy traffic was caused by a terrorist attack at an airport on a nearby island. A bomb explosion at Gran Canaria had injured eight, and a call warning about a second device was received shortly afterward. The authorities decided to temporarily close the facility and divert incoming flights to Los Rodeos. The smaller field didn't have the facilities to cope with all the planes, as it possessed a single runway and a parallel taxiway. The controllers quickly ran out of space and had to park the incoming jets on the taxiway. Some of these were large jets, including Boeing 747s.

The airport's elevation, at 2,000 feet, made it subject to rapidly changing visibility when clouds rolled in. And as the afternoon wore on, a dense fog began to settle over Los Rodeos. This was a problem because the airport lacked ground radar.

Once the all clear had been received from Grand Canaria, the controllers endeavored to get the flights on their way as quickly as possible. Those who had disembarked reboarded the planes. They included the passengers of KLM Flight 4805, a charter from Amsterdam. The Boeing 747 had a crew of 14 and 234 passengers. Another diverted flight was Pan Am 1736, a 747 that had originated at Los Angeles. It had a crew of 16 and 380 passengers. The captains and first officers on both planes had extensive experience flying jumbo jets but were not familiar with Los Rodeos.

Because other diverted aircraft were blocking the taxiway, the ground controllers had the KLM taxi down the entire length of the active runway and then execute a 180-degree turn to prepare for its takeoff. The Los Rodeos ground controllers instructed the Pan Am pilot to follow the KLM jet, take the third runway exit, and move onto the parallel taxiway.

These taxiways, however, were unmarked, and visibility had become extremely poor. Either the pilot missed the third taxiway or decided that the 747 could not negotiate the turns that were required. As a result, the Pan Am stayed on the active runway as it crept forward toward the fourth exit.

The atmospheric conditions at the KLM's end of the runway, however, were much better. Flight 4805 made its turn and began accelerating down the runway, reaching a speed of 160 miles per hour as it approached the Pan Am jet. It had just become airborne and its nose cleared the other plane. Its lower fuselage and the engines, however, ripped through the Pan Am. The KLM jet, which had taken on more fuel at Los Rodeos and was fully loaded, slid down the runway for another 1,000 feet before exploding in flames. By then, the fog was so thick that the fire crews had difficulty finding the crash site. The plane would burn for hours.

Only 61 passengers and crew on the Pan Am flight survived. There were no survivors from the KLM jet. The 583 people who lost their lives died in the deadliest collision in the history of aviation.[43]

The investigation that followed quickly zeroed in on the flight crew of the KLM. They were close to the legal limit for on-duty time. They knew that their flight would be held over at Tenerife if they didn't depart soon.[44]

After reviewing their flight route with the ground controllers, the crew of the KLM had been told, "You are clear." This assent, however, applied only to the route. Even though the pilot knew that a second clearance would be required for takeoff, the flight recorder data show that the plane began to immediately accelerate down the runway.

At this point, the KLM first officer radioed, "We are now at takeoff."

The ground controllers responded with "Okay . . . stand by for takeoff. I will call you."

The Pan Am pilot, at the same time, radioed, "No, uh, we're still taxiing down the runway!"[45]

The flight crew in the KLM cockpit, however, never heard this. Simultaneous radio traffic from the ground controllers and the Pan Am caused interference, and the men heard only a squeal of static.

The copilot of the KLM flight, however, did express concern, saying, "Is he [the Pan Am plane] not clear then?" The pilot, however, simply replied, "Oh, yes," which suggests he thought the runway was clear. The KLM would barrel down the runway until the Pan Am plane became suddenly visible, but by then it was too late to avert a horrific outcome.

In the tragedy at Tenerife, we can see how a perfect storm of miscommunication leads to disaster. Perceptual issues were at play because no one could see what was happening, and the KLM flight crew couldn't hear the Pan Am's frantic radio message. And as we have seen repeatedly, such perceptual difficulties count as a strike against successful communication.

Cognitive issues are also likely to have been contributing factors. The KLM flight crew was fatigued and may have been distracted by time pressure. Even cultural issues may have been involved since the KLM copilot seemed reluctant to forcefully express his concerns to the pilot. Malcolm Gladwell, in his analysis of aircraft accidents, places heavy emphasis on such chain-of-command issues, although they seem to have been only part of the story at Tenerife.[46]

Linguistic code switching played a role as well. The flight crew was conversing in Dutch and then switching to English to speak to the controllers. Differences in the present progressive tense between Dutch and English may have caused the first officer to utter the ambiguous phrase "We are now at takeoff."[47] Did this mean "in position for takeoff," as the Los Rodeos ground controller seems to have thought, or "actually taking off," which is what the flight crew did?

The response of the ground controller was also criticized by the investigators. By replying "Okay"—a nonstandard response—to the first officer's ambiguous statement, the controller seemed to give tacit approval for the takeoff to commence, even though he also instructed the KLM flight crew to stand by.

On a positive note, the tragedy led to changes designed to reduce linguistic ambiguity going forward. As we saw in the previous section, flight crews are now required to read back the critical parts of their instructions as opposed to simply acknowledging them. And the word "takeoff" is now used only when providing actual clearance to fly. All references to that act, before clearance is given, should make use of the word "departure" instead. Such linguistic changes may seem minor, but they have undoubtedly saved lives in the years since that horrific afternoon at Los Rodeos.

EPILOGUE

In general, human communication can be characterized as highly robust and remarkably efficient. We can use language to convey, almost effortlessly, our hopes and our fears. We can call on it to make plans as well as to reflect on the past. Our linguistic abilities allow us to convey subtleties of thought and even to articulate new ideas that can change the world. A particular turn of phrase or even a simple glance or a gesture can be incredibly eloquent and meaningful.

However, as this book has shown, the communicative process can also be described as fairly brittle. Our linguistic abilities may be powerful, but even with the best of intentions, our attempts at communication can be derailed in a variety of ways. The examples I have chosen to include in this book illustrate the wide range of factors that can be at play.

I have also argued that our ability to communicate can tolerate one source of disruption or degradation but typically not more than that. To a large degree, we can characterize the language game as "two strikes and you're out." Perceptual noise, such as in the noisy café we visited in the prologue, counts as one strike against our ability to understand what someone is saying. And if our conversational partner refers to someone we don't know—a second strike—that additional complication can be enough to hobble our ability to understand them.

Our perceptual systems must take a fair share of the blame for our communicative misadventures. Our ability to understand what we hear, read, and see is dependent on complex biological and psychological processes that don't always work the way they're supposed to. For example, we tend to interpret the unfamiliar in terms of the familiar, even when this leads us astray—like Oliver Sacks mishearing "chiropractor" as "choir practice" in chapter 3.

And the world seems to excel at throwing perceptual roadblocks into our paths. Imagine the difficulties that one might experience on a drive to an unfamiliar city. The road signs you encounter might be too small or too hard to read. The roar of highway traffic might make it difficult to hear the directions being rattled off by the car's navigational system. And the unfamiliar accent of someone who has taken pity on you and is providing you with directions can make it difficult to understand them. To make matters worse, such problems are magnified as our perceptual systems age, making it harder to see clearly or to hear well. Given these impediments, it's remarkable that comprehension of what we see and hear is as good as it is.

We have seen throughout this book that communicating by a tweet or a text can easily lead to misunderstanding. And this is because the count against us starts with one strike even before we begin crafting our message. Emails, tweets, and texts are problematic because they lack the sorts of nonverbal cues that language users have relied on for millennia. As a result, our ability to communicate subtleties or nonserious intent is heavily compromised (strike one). And if we don't share common ground or context with the recipient (strike two), such messages often go awry. We saw this in the case of the boy band SB19's tweet in chapter 2 ("Hello, Negros!") and Senator Chuck Grassley's tweet in chapter 5 ("u kno what") as well as in Justine Sacco's edgy tweet about AIDS and white privilege in chapter 9. These communicative misfires underscore how difficult it can be to understand someone when we don't share a context with them and when they broadcast their messages using impoverished media like Twitter or Facebook.

But our ability to understand others isn't undermined only by computer-mediated communication. As we saw in chapter 6, unfamiliar or ambiguous gestures or hand signals can also be a source of confusion. Even silence can be misinterpreted.

In addition, the language itself deserves its share of blame. A tongue like English sometimes seems to have been designed by a committee whose members had their own share of communication failures. As we saw in chapter 4, some sounds are hard to pronounce, and some strings of sounds map on to more than one concept. The meanings of words can be ambiguous, as we saw in chapter 7, and this can be true for sentences as well. Even our punctuation can let us down. English can soar to great heights in prose and in poetry, but as a tool for communication, it can sometimes be maddeningly vague or imprecise.

And since language is a social instrument, the role of society must be considered as well. Human cultures are constantly evolving and changing, and language has to adapt to these changes. New words seem to pop up, seemingly out of thin air, while older terms fall out of favor or mark their users as out of step with the times.

Perhaps even more confusingly, some terms can suddenly shift in their meaning, as we saw in chapter 5. A completely innocent name like "Karen," for example, was transformed, seemingly overnight, into a pejorative and dismissive term for a middle-aged, privileged white woman who always demands to speak to the manager. And the use of words or phrases that were once unobjectionable can be perceived as racist, ageist, or sexist.

Miscommunication is also a consequence of how our minds work. As we saw in chapter 1, our expectations can sometimes lead us astray, or our frame of reference may not align with that of our audience. One of the most pernicious of these factors may be our egocentrism. On the one hand, it's difficult to imagine *not* viewing the world from one's own perspective. And in terms of personal survival, a certain degree of self-centeredness is probably highly adaptive. But egocentrism can give rise to a host of issues that impair communication. Several consequences of egocentrism were described in chapter 2, including the curse of knowledge. Because it is difficult for us to imagine not knowing something, we may tend to under-explain a procedure that is well known to us or to assume that our sarcasm or nonserious intent will be easily understood by others.

Egocentrism can also lead to failures in monitoring the common ground that we share with others. Grounding new terms and concepts for our listeners and readers can be cognitively taxing. And keeping track of who knows what can be mentally demanding as well. As a result, when we lack sufficient bandwidth to track this shared common ground, we may leave others confused or unclear about who or what we're referring to.

One reason that I wrote a book about failures of communication was to highlight their common elements. Many communicative misadventures have similarities and connections that may not be apparent at first blush. But on closer inspection, many types of misunderstandings, even those that seem very different from one another, have a similar underlying cause.

To consider but one example, online flame wars and road rage appear to have little in common. However, both may be the result of impoverished communication channels. Our inability to clearly mark humor, sarcasm, and innuendo in social media posts may inflame others who fail to understand our intent. And our inability to adequately signal contrition to other drivers may cause them to assume the worst about our perceptual abilities and driving skills.

Another example would be the "broken telephone" effects discussed at several points throughout the book. These can be seen in many domains that go far beyond the simple game played by children. Long chains of communication can help account for the distorting effects of gossip, described in chapter 8, as well as misadventure on the battlefield, illustrated in chapter 10.

So is there any hope for us? Are we condemned to lives full of confusion, mistakes, and misunderstanding? I do have one concrete suggestion to offer. If we find ourselves in situations where there is already one communicative strike working against us, we should take that into account and try to compensate in some way. For example, a dinner with a hard-of-hearing parent at a noisy restaurant would probably not be the best place to discuss a complex financial transaction. Similarly, we should realize that edgy social media posts will almost inevitably be misinterpreted by someone, somewhere, at some point in time. And our attempts at being sarcastic or humorous in an email should probably be garnished with a healthy dollop of emoticons, emoji, or other markers of nonliteral intent.

Finally, an important goal of mine has been to share the results of the ingenious studies devised by psychologists and other researchers who study communication. A great deal has been learned, but some of the factors involved in miscommunication have yet to be described or are still poorly understood. It is my hope that in the coming years, researchers will be able to fill in the gaps to create a more complete picture of why we misunderstand what we hear, read, and see.

NOTES

CHAPTER 1

1. Brian Knowlton and the International Herald Tribune, "No Radio Contact with NASA Spacecraft: Orbiter Believed Lost upon Arrival at Mars," *New York Times*, September 24, 1999.

2. Beth Dickey, "Spacecraft Is Launched to Look for Water on Mars," *New York Times*, December 12, 1998, A10.

3. John N. Wilford, "Mars Orbiting Craft Presumed Destroyed by Navigation Error," *New York Times*, September 24, 1999, A01.

4. Larry J. Paxton, "'Faster, Better, and Cheaper' at NASA: Lessons Learned in Managing and Accepting Risk," *Acta Astronautica* 61 (2007): 954–63.

5. Peter Francis Jr., "The Beads That Did 'Not' Buy Manhattan Island," *New York History* 78, no. 4 (October 1997): 411–28.

6. Bruce Shentiz, "New York's Beginnings, Real and Imagined," *New York Times*, December 3, 1999, E35.

7. Jason Barr, Fred H. Smith, and Sayali J. Kulkarni, "What's Manhattan Worth? A Land Values Index from 1950 to 2014," *Regional Science and Urban Economics* 70 (2018): 1–19.

8. Edward Robb Ellis, *The Epic of New York City: A Narrative History* (New York: Basic Books, 1966).

9. Dave Itzkoff, "Tony Lives, or Doesn't," *New York Times*, August 29, 2014, C3.

10. Frederic C. Bartlett, *Remembering: A Study in Experimental and Social Psychology* (Cambridge: Cambridge University Press, 1932/1997).

11. William F. Brewer, "Bartlett's Concept of the Schema and Its Impact on Theories of Knowledge Representation in Contemporary Cognitive Psychology," in *Bartlett, Culture and Cognition*, ed. Akiko Saito (Hove: Psychology Press, 2000), 69–89.

12. Robert P. Abelson, "Psychological Status of the Script Concept," *American Psychologist* 36, no. 7 (1981): 715–29.

13. Jon Morgan, "Radar, Built Here, Detected Pearl Harbor Attack, but . . . Futile Early Warning," *Baltimore Sun*, November 20, 1991, https://www.baltimoresun.com/news/bs-xpm-1991-11-29-1991333114-story.html.

14. Brenda Wineapple, *The Impeachers: The Trial of Andrew Johnson and the Dream of a Just Nation* (New York: Random House, 2019).

15. Daniel M. Goldberg, "What Comprises a 'Lascivious Exhibition of the Genitals or Pubic Area'? The Answer, My Friend, Is *Blouin* in the Wind," *Military Law Review* 224 (2016): 425–80.

16. Lara N. Strayer, "Ambiguous Laws Do Little to Erase Kiddie Porn," *Temple Political and Civil Rights Law Review* 5 (1995): 169–81.

17. Ryan McCarl, "Incoherent and Indefensible: An Interdisciplinary Critique of the Supreme Court's Void-for-Vagueness Doctrine," *Hastings Constitutional Law Quarterly* 42 (2014): 73–94.

18. Risa Goluboff, "The Forgotten Law That Gave Police Nearly Unlimited Power," *Time*, February 1, 2016.

19. John Pratt and Michelle Miao, "Risk, Populism, and Criminal Law," *New Criminal Law Review* 22, no. 4 (2019): 391–433.

20. James Madison, "Concerning the Difficulties of the Convention in Devising a Proper Form of Government," *Federalist Paper No. 37*, 1788.

21. David Dorsen, "Is Gorsuch an Originalist? Not So Fast," *Washington Post*, March 17, 2017.

22. Margaret Jane Radin, *Boilerplate: The Fine Print, Vanishing Rights, and the Rule of Law* (Princeton, NJ: Princeton University Press, 2012).

23. Yannis Bakos, Florencia Marotta-Wurgler, and David R. Trossen, "Does Anyone Read the Fine Print? Consumer Attention to Standard-Form Contracts," *Journal of Legal Studies* 43, no. 1 (2014): 1–35.

24. *Bartlett's Book of Love Quotations* (New York: Little, Brown, 1994).

25. Khalid Khulaif Alshammari, "Directness and Indirectness of Speech Acts in Requests among American Native English Speakers and Saudi Native Speakers of Arabic," *English Literature and Language Review* 1, no. 8 (2015): 63–69.

26. Eva Alcón Soler, "Does Instruction Work for Learning Pragmatics in the EFL Context?" *System* 33, no. 3 (2005): 417–35.

27. Christopher Borrelli, "Reagan Used Her, the Country Hated Her. Decades Later, the Welfare Queen of Chicago Refuses to Go Away," *Chicago Tribune*, June 10, 2019.

28. Rachel Wetts and Robb Willer, "Who Is Called by the Dog Whistle? Experimental Evidence That Racial Resentment and Political Ideology Condition Responses to Racially Encoded Messages," *Socius* 5 (2019): 1–20.

29. Michael Kazin, *The Populist Persuasion: An American History* (Ithaca, NY: Cornell University Press, 1998).

30. David D. Kirkpatrick, "Speaking in the Tongue of Evangelicals," *New York Times*, October 17, 2004.

31. Ben Sales, "Senator's Speech on 'Cosmopolitan Elites': Anti-Semitic Dog Whistle or Poli-Sci Speak?" *Jewish Telegraphic Agency*, July 19, 2019, https://www.jta.org/2019/07/19/united-states/a-missouri-senator-gave-a-speech-opposing-a-powerful-upper-class-and-their-cosmopolitan-priorities-um.

32. Jason Hancock, "Hawley's Critique of 'Cosmopolitan Elite' Earns Rebuke from Missouri Jewish Leaders," *Kansas City Star*, July 19, 2019, https://www.kansascity.com/news/politics-government/article232902747.html.

33. Ian Olasov, "Offensive Political Dog Whistles: You Know Them When You Hear Them. Or Do You?" *Vox*, November 7, 2016, https://www.vox.com/the-big-idea/2016/11/7/13549154/dog-whistles-campaign-racism.

34. Herbert A. Simon, "Rational Choice and the Structure of the Environment," *Psychological Review* 63, no. 2 (1956): 129–38.

35. H. P. Grice, "Logic and Conversation," in *Syntax and Semantics III: Speech Acts*, ed. P. Cole and J. L. Morgan (New York: Academic Press, 1975), 41–58.

36. Paul E. Engelhardt, Karl G. D. Bailey, and Fernanda Ferreira, "Do Speakers and Listeners Observe the Gricean Maxim of Quantity?" *Journal of Memory and Language* 54, no. 4 (2006): 554–73.

37. Paula Rubio-Fernandez, "Overinformative Speakers Are Cooperative: Revisiting the Gricean Maxim of Quantity," *Cognitive Science* 43, no. 11 (2019): e12797.

38. Jean-Marc Dewaele, "Interpreting Grice's Maxim of Quantity: Interindividual and Situational Variation in Discourse Styles of Non Native Speakers," in *Cognition in Language Use: Selected Papers from the 7th International Pragmatics Conference*, Vol. 1, ed. E. Németh (Antwerp: International Pragmatics Association, 2001), 85–99.

39. Warren St. John, "When Information Becomes T.M.I.," *New York Times*, September 10, 2006, sec. 9.

40. Kevin Finneran, "Can the Public Be Trusted?" *Issues in Science and Technology* 34, no. 4 (2018): 17–18.

41. Murat Akçayır, Hakan Dündar, and Gökçe Akçayır, "What Makes You a Digital Native? Is It Enough to Be Born after 1980?" *Computers in Human Behavior* 60 (2016): 435–40.

42. Paul A. Kirschner and Pedro De Bruyckere, "The Myths of the Digital Native and the Multitasker," *Teaching and Teacher Education* 67 (2017): 135–42.

43. Gretchen McCulloch, *Because Internet: Understanding the New Rules of Language* (New York: Riverhead Books, 2019).

44. Keri K. Stephens, Marian L. Houser, and Renee L. Cowan, "RU Able to Meat Me: The Impact of Students' Overly Casual Email Messages to Instructors," *Communication Education* 58, no. 3 (2009): 303–26.

45. Danielle N. Gunraj, April M. Drumm-Hewitt, Erica M. Dashow, Sri Siddhi N. Upadhyay, and Celia M. Klin, "Texting Insincerely: The Role of the Period in Text Messaging," *Computers in Human Behavior* 55 (2016): 1067–75.

46. Kenneth J. Houghton, Sri Siddhi N. Upadhyay, and Celia M. Klin, "Punctuation in Text Messages May Convey Abruptness. Period," *Computers in Human Behavior* 80 (2018): 112–21.

47. Jessica Bennett, "When Your Punctuation Says It All (!)," *New York Times*, March 1, 2015, ST2.

48. Logan Mahan, "Youthsplaining: You've Been Texting the Word 'Okay' Wrong," *InsideHook*, July 31, 2019, https://www.insidehook.com/article/advice/the-difference-between-texting-k-ok-kk-explained.

CHAPTER 2

1. Elizabeth L. Newton, "The Rocky Road from Actions to Intentions," unpublished doctoral dissertation, Stanford University, Stanford, CA, 1991.

2. William H. Whyte, "Is Anybody Listening?" *Fortune*, September 1950.

3. Boaz Keysar, "Communication and Miscommunication: The Role of Egocentric Processes," *Intercultural Pragmatics*, 4-1 (2007): 71-84.

4. Lee Ross, David Greene, and Pamela House, "The 'False Consensus Effect': An Egocentric Bias in Social Perception and Attribution Processes," *Journal of Experimental Social Psychology* 13, no. 3 (1977): 279–301.

5. Thomas Gilovich, Victoria Husted Medvec, and Kenneth Savitsky, "The Spotlight Effect in Social Judgment: An Egocentric Bias in Estimates of the Salience of One's Own Actions and Appearance," *Journal of Personality and Social Psychology* 78, no. 2 (2000): 211–22.

6. Thomas Gilovich, Kenneth Savitsky, and Victoria Husted Medvec, "The Illusion of Transparency: Biased Assessments of Others' Ability to Read One's Emotional States," *Journal of Personality and Social Psychology* 75, no. 2 (1998): 332–46.

7. Roshan Rai, Peter Mitchell, Tasleem Kadar, and Laura Mackenzie, "Adolescent Egocentrism and the Illusion of Transparency: Are Adolescents as Egocentric as We Might Think?" *Current Psychology* 35, no. 3 (2016): 285–94.

8. Emily Pronin, Justin Kruger, Kenneth Savtisky, and Lee Ross, "You Don't Know Me, but I Know You: The Illusion of Asymmetric Insight," *Journal of Personality and Social Psychology* 81, no. 4 (2001): 639–56.

9. Boaz Keysar and Anne S. Henly, "Speakers' Overestimation of Their Effectiveness," *Psychological Science* 13, no. 3 (2002): 207–12.

10. Michael F. Schober and Herbert H. Clark, "Understanding by Addressees and Overhearers," *Cognitive Psychology* 21, no. 2 (1989): 211–32.

11. Herbert H. Clark and Susan E. Brennan, "Grounding in Communication," in *Perspectives on Socially Shared Cognition*, ed. Lauren B. Resnick, John M. Levine, and Stephanie D. Teasley (Washington DC: American Psychological Association, 1991), 127–49.

12. Herbert H Clark, *Using Language* (Cambridge: Cambridge University Press, 1996).

13. Edda Weigand, "Misunderstanding: The Standard Case," *Journal of Pragmatics* 31, no. 6 (1999): 763–85.

14. Ashwin Rajadesingan, Reza Zafarani, and Huan Liu, "Sarcasm Detection on Twitter: A Behavioral Modeling Approach," in *Proceedings of the Eighth ACM International Conference on Web Search and Data Mining*, ed. Xueqi Cheng (New York: Association for Computing Machinery, 2015), 97–106.

15. Roger Kreuz, *Irony and Sarcasm* (Cambridge, MA: MIT Press, 2020).

16. Jean E. Fox Tree, J. Trevor D'Arcey, Alicia A. Hammond, and Alina S. Larson, "The Sarchasm: Sarcasm Production and Identification in Spontaneous Conversation," *Discourse Processes* 2020: 1–27.

17. Stacey L. Ivanko, Penny M. Pexman, and Kara M. Olineck, "How Sarcastic Are You? Individual Differences and Verbal Irony," *Journal of Language and Social Psychology* 23, no. 3 (2004): 244–71.

18. Gina M. Caucci and Roger J. Kreuz, "Social and Paralinguistic Cues to Sarcasm," *Humor* 25, no. 1 (2012): 1–22.

19. Raymond W. Gibbs Jr., "Irony in Talk among Friends," *Metaphor and Symbol* 15, no. 1–2 (2000): 5–27.

20. Roger J. Kreuz and Richard M. Roberts, "Two Cues for Verbal Irony: Hyperbole and the Ironic Tone of Voice," *Metaphor and Symbol* 10, no. 1 (1995): 21–31.

21. Soujanya Poria, Devamanyu Hazarika, Navonil Majumder, and Rada Mihalcea, "Beneath the Tip of the Iceberg: Current Challenges and New Directions in Sentiment Analysis Research," preprint, arXiv:2005.00357 (2020).

22. Cynthia Van Hee, Els Lefever, and Véronique Hoste, "We Usually Don't Like Going to the Dentist: Using Common Sense to Detect Irony on Twitter," *Computational Linguistics* 44, no. 4 (2018): 793–832.

23. Pagan Kennedy, "Who Made That Emoticon?" *New York Times*, November 23, 2012.

24. Debanjan Ghosh, Alexander R. Fabbri, and Smaranda Muresan, "Sarcasm Analysis Using Conversation Context," *Computational Linguistics* 44, no. 4 (2018): 755–92.

25. John Markoff, "Computer Wins on 'Jeopardy!': Trivial, It's Not," *New York Times*, February 16, 2011.

26. Rob High and Tanmay Bakshi, *Cognitive Computing with IBM Watson* (Birmingham: Packt Publishing, 2019).

27. *Nova*, "Smartest Machine on Earth," directed by Michael Bicks, written by Michael Bicks and Julia Cort, *PBS*, February 9, 2011.

28. Melanie Mitchell, *Artificial Intelligence: A Guide for Thinking Humans* (New York: Farrar, Straus and Giroux, 2019).

29. Van Hee et al., "We Usually Don't Like Going to the Dentist."

30. Cade Metz, "One Genius' Lonely Crusade to Teach a Computer Common Sense," *Wired*, March 24, 2016.

31. Thomas D. Erickson and Mark E. Mattson, "From Words to Meaning: A Semantic Illusion," *Journal of Verbal Learning and Verbal Behavior* 20, no. 5 (1981): 540–51.

32. Craig Speelman, "Implicit Expertise: Do We Expect Too Much from Our Experts?" in *Implicit and Explicit Mental Processes*, ed. Kim Kirsner, Craig Speeman, Murray Mabery, Angela O'Brien-Malone, Mike Anderson, and Colin MacLeod (Mahwah, NJ: Lawrence Erlbaum Associates, 1998), 135–47.

33. Allison D. Cantor and Elizabeth J. Marsh, "Expertise Effects in the Moses Illusion: Detecting Contradictions with Stored Knowledge," *Memory* 25, no. 2 (2017): 220–30.

34. Krissy Aguilar, "SB19 Accused of 'Racism' after Tweeting 'Hello, Negros!' for Concert Tour," *Inquirer.net*, December 27, 2019, https://

entertainment.inquirer.net/356469/sb19-accused-of-racism-after-tweeting -hello-negros-for-concert-tour.

35. Rich Duprey, "Drinkers Are Confusing Corona Beer with the Coronavirus," *The Motley Fool*, February 5, 2020, https://www.fool.com/investing/2020/02/05/drinkers-are-confusing-corona-beer-with-the-corona.aspx.

36. Kelly Tyko, "'Corona Beer Virus' and 'Beer Coronavirus' Searches Increase as Fears of Outbreak Spread," *USA Today*, February 27, 2020.

37. Michael M. Ego, "'Chink in the Armor': Is It a Racist Cliché?" *AP News*, October 28, 2018, https://apnews.com/4722f93c10a141d39ddb0a65a ef16c49.

38. Irving DeJohn and Helen Kennedy, "Jeremy Lin Headline Slur Was 'Honest Mistake,' Fired ESPN Editor Anthony Federico Claims," *Daily News*, February 20, 2012. https://www.nydailynews.com/sports/basketball/knicks/ jeremy-lin-slur-honest-mistake-fired-espn-editor-anthony-federico-claims -article-1.1025566.

39. Richard Sandomir, "ESPN Fires Employee Over Slur," *New York Times*, February 20, 2012, D2.

40. Robert E Haskell, "Unconscious Linguistic Referents to Race: Analysis and Methodological Frameworks," *Discourse and Society* 20, no. 1 (2009): 59–84.

41. Sigmund Freud, *The Psychopathology of Everyday Life* (New York: Norton, 1901/1989).

42. Jerry Gary, "No. 2 House Leader Refers to Colleague with Anti-Gay Slur," *New York Times*, January 28, 1995, 1.

43. Saul Albert and J. P. De Ruiter, "Repair: The Interface between Interaction and Cognition," *Topics in Cognitive Science* 10, no. 2 (2018): 279–313.

44. Emanuel A. Schegloff, Gail Jefferson, and Harvey Sacks, "The Preference for Self-Correction in the Organization of Repair in Conversation," *Language* 53, no. 2 (1977): 361–82.

45. Willem J. M. Levelt, "Monitoring and Self-Repair in Speech," *Cognition* 14, no. 1 (1983): 41–104.

46. Mark Dingemanse, Seán G. Roberts, Julija Baranova, Joe Blythe, Paul Drew, Simeon Floyd, Rosa S. Gisladottir, Kobin H. Kendrick, Stephen C. Levinson, Elizabeth Manrique, Giovanni Rossi, and N. J. Enfield, "Universal Principles in the Repair of Communication Problems," *PLoS One* 10, no. 9 (2015): e0136100.

47. Patrick G. T. Healey, Gregory J. Mills, Arash Eshghi, and Christine Howes, "Running Repairs: Coordinating Meaning in Dialogue," *Topics in Cognitive Science* 10, no. 2 (2018): 367–88.

48. Simeon Floyd, Elizabeth Manrique, Giovanni Rossi, and Francisco Tor-reira, "Timing of Visual Bodily Behavior in Repair Sequences: Evidence from Three Languages," *Discourse Processes* 53, no. 3 (2016): 175.

49. *Seinfeld*, "The Subway," episode 313, directed by Tom Cherones, writ-ten by Larry Charles, NBC, January 8, 1992.

50. Schegloff et al., "The Preference for Self-Correction in the Organization of Repair in Conversation."

CHAPTER 3

1. Scott Brown, "Daily Poll: Do You Hear Laurel or Yanny?" *Vancouver Sun*, May 16, 2018.

2. Matteo Toscani, Karl R. Gegenfurtner, and Katja Doerschner, "Dif-ferences in Illumination Estimation in #thedress," *Journal of Vision* 17, no. 1 (2017): 1–14.

3. Josh Katz, Jonathan Corum, and Jon Huang, "We Made a Tool So You Can Hear Both Yanny and Laurel," *New York Times*, May 16, 2018.

4. Paul Vitello, "A Ring Tone Meant to Fall on Deaf Ears," *New York Times*, June 12, 2006.

5. Mitchell Akiyama, "Silent Alarm: The Mosquito Youth Deterrent and the Politics of Frequency," *Canadian Journal of Communication* 35 (2010): 455–71.

6. Jerry L. Northern and Marion P. Downs, *Hearing in Children*, 5th ed. (Philadelphia: Lippincott Williams & Wilkins, 2002).

7. Piers Dawes, Richard Emsley, Karen J. Cruickshanks, David R. Moore, Heather Fortnum, Mark Edmondson-Jones, Abby McCormack, and Kevin J. Munro, "Hearing Loss and Cognition: The Role of Hearing Aids, Social Isola-tion and Depression," *PLoS One* 10, no. 3 (2015): e0119616.

8. Susan Kemper, "Elderspeak: Speech Accommodations to Older Adults," *Aging and Cognition* 1, no. 1 (1994): 17–28.

9. David M. Harland, *Apollo 12—On the Ocean of Storms* (New York: Springer, 2011); Rick Houston and Milt Heflin, *Go Flight! The Unsung Heroes of Mission Control, 1965–1992* (Lincoln: University of Nebraska Press, 2015).

10. Frédéric Apoux and Sid P. Bacon, "Relative Importance of Temporal Information in Various Frequency Regions for Consonant Identification in Quiet and in Noise," *Journal of the Acoustical Society of America* 116, no. 3 (2004): 1671–80.

11. R. Gary Patterson, *The Walrus Was Paul: The Great Beatle Death Clues* (New York: Fireside, 1996).

12. Jonathan Cott, "John Lennon: The Rolling Stone Interview," *Rolling Stone*, November 23, 1968.

13. Jiangang Liu, Jun Li, Lu Feng, Ling Li, Jie Tian, and Kang Lee, "Seeing Jesus in Toast: Neural and Behavioral Correlates of Face Pareidolia," *Cortex* 53 (2014): 60–77.

14. John McGrath, Sukanta Saha, David Chant, and Joy Welham, "Schizophrenia: A Concise Overview of Incidence, Prevalence, and Mortality," *Epidemiologic Reviews* 30, no. 1 (2008): 67–76.

15. K. Maijer, M. J. H. Begemann, S. J. M. C. Palmen, S. Leucht, and I. E. C. Sommer, "Auditory Hallucinations across the Lifespan: A Systematic Review and Meta-analysis," *Psychological Medicine* 48, no. 6 (2018): 879–88.

16. Ben Alderson-Day, Cesar F. Lima, Samuel Evans, Saloni Krishnan, Pradheep Shanmugalingam, Charles Fernyhough, and Sophie K. Scott, "Distinct Processing of Ambiguous Speech in People with Non-Clinical Auditory Verbal Hallucinations," *Brain* 140, no. 9 (2017): 2475–89.

17. Judith M. Ford, Thomas Dierks, Derek J. Fisher, Christoph S. Herrmann, Daniela Hubl, Jochen Kindler, Thomas Koenig, Daniel H. Mathalon, Kevin M. Spencer, Werner Strik, and Remko van Lutterveld, "Neurophysiological Studies of Auditory Verbal Hallucinations," *Schizophrenia Bulletin* 38, no. 4 (2012): 715–23.

18. Neville Moray, "Attention in Dichotic Listening: Affective Cues and the Influence of Instructions," *Quarterly Journal of Experimental Psychology* 11, no. 1 (1959): 56–60.

19. Oliver Sacks, *Hallucinations* (New York: Vintage Books, 2012).

20. S. F. Crowe, J. Barot, S. Caldow, J. d'Aspromonte, J. Dell'Orso, A. Di Clemente, K. Hanson, M. Kellett, S. Makhlota, B. McIvor, L. McKenzie, R. Norman, A. Thiru, M. Twyerould, and S. Sapega, "The Effect of Caffeine and Stress on Auditory Hallucinations in a Non-Clinical Sample," *Personality and Individual Differences* 50, no. 5 (2011): 626–30.

21. Susan Wright, "The Death of Lady Mondegreen," *Harper's Magazine* 201 (November 1954): 48–51.

22. Dominic W. Massaro and Alexandra Jesse, "Read My Lips: Speech Distortions in Musical Lyrics Can Be Overcome (Slightly) by Facial Information," *Speech Communication* 51, no. 7 (2009): 604–21.

23. Susan E. Brennan, Marie K. Huffman, and S. Hannigan, "More Adventures with Dialects: Convergence and Ambiguity Resolution by Partners in Conversation" (n.d.), https://slideplayer.com/slide/4018652.

24. Gerritt Kentner, "Rhythmic Segmentation in Auditory Illusions: Evidence from Cross-Linguistic Mondegreens," in *Proceedings of the 18th International Congress of Phonetic Sciences* (Glasgow: University of Glasgow, 2015).

25. Oliver Sacks, "Mishearings," *New York Times,* June 6, 2015, SR4.

26. J. Don Read, "Detection of Fs in a Single Statement: The Role of Phonetic Recoding," *Memory and Cognition* 11, no. 4 (1983): 390–99.

27. Peter F. De Jong, Daniëlle J. L. Bitter, Margot Van Setten, and Eva Marinus, "Does Phonological Recoding Occur during Silent Reading, and Is It Necessary for Orthographic Learning?" *Journal of Experimental Child Psychology* 104, no. 3 (2009): 267–82.

28. Keith Rayner, "Eye Movements in Reading and Information Processing: 20 Years of Research," *Psychological Bulletin* 124, no. 3 (1998): 372–422.

29. Edward J. O'Brien, Anne E. Cook, and Robert F. Lorch Jr., *Inferences during Reading* (Cambridge: Cambridge University Press, 2015).

30. Joel Sherzer, *Speech Play and Verbal Art* (Austin: University of Texas Press, 2002).

31. Adam Drewnowski and Alice F. Healy, "Detection Errors on *the* and *and*: Evidence for Reading Units Larger Than the Word," *Memory and Cognition* 5, no. 6 (1977): 636–47.

32. Adrian Staub, Sophia Dodge, and Andrew L. Cohen, "Failure to Detect Function Word Repetitions and Omissions in Reading: Are Eye Movements to Blame?" *Psychonomic Bulletin and Review* 26, no. 1 (2019): 340–46.

33. Meredyth Daneman and Murray Stainton, "The Generation Effect in Reading and Proofreading: Is It Easier or Harder to Detect Errors in One's Own Writing?" *Reading and Writing* 5, no. 3 (1993): 297–313.

34. Jeremy Caplan, "Cause of Death: Sloppy Doctors," *Time*, May 15, 2007, http://content.time.com/time/health/article/0,8599,1578074,00.html.

35. Leonardo L. Leonidas, "Death by Bad Handwriting," *Philippine Daily Inquirer*, October 27, 2014, https://opinion.inquirer.net/79623/death-by-bad-handwriting.

36. Adrienne Berman, "Reducing Medication Errors through Naming, Labeling, and Packaging," *Journal of Medical Systems* 28, no. 1 (2004): 9–29.

37. David C. Radley, Melanie R. Wasserman, Lauren E. W. Olsho, Sarah J. Shoemaker, Mark D. Spranca, and Bethany Bradshaw, "Reduction in Medication Errors in Hospitals Due to Adoption of Computerized Provider Order Entry Systems," *Journal of the American Medical Informatics Association* 20, no. 3 (2013): 470–76.

38. Emily S. Rueb, "Cursive Seemed to Go the Way of Quills and Parchment. Now It's Coming Back," *New York Times*, April 13, 2019, https://www.nytimes.com/2019/04/13/education/cursive-writing.html.

39. Anne Trubek, *The History and Uncertain Future of Handwriting* (New York: Bloomsbury, 2016).

40. Joe Heim, "Once All but Left for Dead, Is Cursive Handwriting Making a Comeback?" *Washington Post*, July 26, 2016.

41. Ron Nixon, "Last of a Breed: Postal Workers Who Decipher Bad Addresses," *New York Times*, May 4, 2013, A11.

42. Russell W. Driver, M. Ronald Buckley, and Dwight D. Frink, "Should We Write Off Graphology?" *International Journal of Selection and Assessment* 4, no. 2 (1996): 78–86.

43. Helmut Ploog, *Handwriting Psychology: Personality Reflected in Handwriting* (Bloomington, IN: iUniverse, 2013), 89.

44. Carla Dazzi and Luigi Pedrabissi, "Graphology and Personality: An Empirical Study on Validity of Handwriting Analysis," *Psychological Reports* 105, no. 3, pt. 2 (2009): 1255–68.

45. Sara Rosenblum, Margalit Samuel, Sharon Zlotnik, Ilana Erikh, and Ilana Schlesinger, "Handwriting as an Objective Tool for Parkinson's Disease Diagnosis," *Journal of Neurology* 260, no. 9 (2013): 2357–61.

CHAPTER 4

1. Marc Fisher, "Donald Trump Doesn't Read Much. Being President Probably Wouldn't Change That," *Washington Post*, July 17, 2016.

2. Thomas Putnam, "The Real Meaning of *Ich Bin ein Berliner*," *The Atlantic*, August 2013.

3. Kat Eschner, "Where the Myth of JFK's 'Jelly Donut' Mistake Came From," *Smithsonian Magazine*, June 26, 2017.

4. Mariel Padilla, "Wait, Is It Nevada, or Nevada?" *New York Times*, February 19, 2020.

5. Olivia Waxman, "Pronunciation Fail Costs Guy $1 Million Prize on 'Wheel of Fortune,'" *Time*, September 19, 2013.

6. Mary L. Roberts, *What Soldiers Do: Sex and the American GI in World War II France* (Chicago: University of Chicago Press, 2013).

7. Chengsheng Li, Ruilan Lu, and D. Li Davis, "Let 500,000,000 Bottles Be Broken All Over China: Underground Dissidents Are Hoping the Small

Gesture Will Breathe New Life into the Pro-Democracy Movement," *Los Angeles Post*, April 18, 1990.

8. Jack Rosenthal, "On Language: Misheard, Misread, Misspoken," *New York Times*, 20, August 28, 1988.

9. Mark Liberman, "Egg Corns: Folk Etymology, Malapropism, Mondegreen, ???" *Language Log*, September 23, 2003, http://itre.cis.upenn.edu/~myl/languagelog/archives/000018.html.

10. Mark Memmott, "'Eggcorns': The Gaffes That Spread Like Wildflowers," *NPR Weekend Edition Saturday*, May 30, 2015, https://www.npr.org/sections/thetwo-way/2015/05/30/410504851/eggcorns-the-gaffes-that-spread-like-wildflowers.

11. James. J. Kilpatrick, *The Ear Is Human: A Handbook of Homophones and Other Confusions* (Kansas City, MO: Andrews, McMeel & Parker, 1985).

12. Bill Bryson, *The Mother Tongue: English and How It Got That Way* (New York: W. Morrow, 1990).

13. Richard Lederer, "Curious Contronyms," *Word Ways*, February 1978, 27–28.

14. Merriam-Webster, "Did We Change the Definition of Literally?" https://www.merriam-webster.com/words-at-play/misuse-of-literally.

15. Benjamin Dreyer, *Dreyer's English: An Utterly Correct Guide to Clarity and Style* (New York: Random House, 2019), 163.

16. Brad Leithauser, "Unusable Words," *The New Yorker*, October 14, 2013.

17. Michael S. Wogalter, Jesseca R. Israel, Soyun Kim, Emily R. Morgan, Kwamoore M. Coleman, and Julianne West, "Hazard Connotation of Fire Safety Terms," in *Proceedings of the Human Factors and Ergonomics Society Annual Meeting* 54, no. 21 (Thousand Oaks CA: Sage Publications, 2010), 1837–40.

18. Patricia T. O'Conner and Stewart Kellerman, *Origins of the Specious: Myths and Misconceptions of the English Language* (New York: Random House, 2009).

19. "'In' on 'Inflammable' Found to Mislead Many," *New York Times*, January 21, 1947, 10.

20. Keith Allan, "The Pragmatics of Connotation," *Journal of Pragmatics* 39, no. 6 (2007): 1047–57.

21. Abigail Herbst, "How the Term 'Social Justice Warrior' Became an Insult," *Foundation for Economic Education*, August 13, 2018, https://fee.org/articles/how-the-term-social-justice-warrior-became-an-insult.

22. Nick Cumming-Bruce, "Myanmar Generals Should Face Genocide Charges Over Rohingya, U.N. Says," *New York Times*, August 27, 2018.

23. Nick Cummings-Bruce, "U.S. Steps Up Criticism of China for Detentions in Xinjiang," *New York Times*, March 13, 2019.

24. "Words Came in Marked for Death . . . ," *The New Yorker*, April 23, 2012.

25. Paul H. Thibodeau, "A Moist Crevice for Word Aversion: In Semantics Not Sounds," *PLoS One* 11.4 (2016): e0153686.

26. "Romney Asserts He Underwent 'Brainwashing' on Vietnam Trip," *New York Times*, September 5, 1967, 28.

27. Merrill Perlman, "How the Word 'Queer' Was Adopted by the LGBTQ Community," *Columbia Journalism Review*, February 22, 2019, https://www.cjr.org/language_corner/queer.php.

28. Hy Harrison, "The Origin of 'Yankee,'" *The Academy and Literature* (1913): 222–23.

29. Peter McNeil, "'That Doubtful Gender': Macaroni Dress and Male Sexualities," *Fashion Theory* 3, no. 4 (1999): 411–47.

30. Allan M. Siegal and William G. Connolly, *The New York Times Manual of Style and Usage*, 5th ed. (New York: Three Rivers Press, 2015).

31. Bryan Garner, *Garner's Modern English Usage*, 4th ed. (Oxford: Oxford University Press, 2016), 114.

32. Merriam-Webster, "Bimonthly," https://www.merriamwebster.com/dictionary/bimonthly.

33. Garner, *Garner's Modern English Usage.*

34. Anke Holler and Katja Suckow, eds., *Empirical Perspectives on Anaphora Resolution* (Berlin: Walter de Gruyter, 2016).

35. Massimo Poesio, Roland Stuckardt, and Yannick Versley, eds., *Anaphora Resolution: Algorithms, Resources, and Applications* (Berlin: Springer-Verlag, 2016).

36. Nicolas Nicolov, "Book Review: Anaphora Resolution," *IEEE Computational Intelligence Bulletin* 2, no. 1 (2003): 31–32.

37. Morton Ann Gernsbacher and David J. Hargreaves, "Accessing Sentence Participants: The Advantage of First Mention," *Journal of Memory and Language* 27, no. 6 (1988): 699–717.

38. Barbara Hemforth, Lars Konieczny, Christoph Scheepers, Saveria Colonna, Sarah Schimke, Peter Baumann, and Joël Pynte, "Language Specific Preferences in Anaphor Resolution: Exposure or Gricean Maxims?" in *32nd Annual Conference of the Cognitive Science Society* 2010: 2218–23, hal-01015048.

39. Eric Topol, *The Creative Destruction of Medicine: How the Digital Revolution Will Create Better Health Care* (New York: Basic Books, 2012).

40. Tom Goldstein, "Lawyers Now Confuse Even the Same Aforementioned," *New York Times*, April 1, 1977, 23.

41. Robert W. Benson, "The End of Legalese: The Game Is Over," *NYU Review of Law and Social Change* 13 (1984): 519.

42. Martin A. Schwartz, "Legal Writing: Legalese, Please," *Probate and Property* 31 (2017): 55–59.

43. Michael Stephenson, "Harry Potter Language: The Plain Language Movement and the Case against Abandoning Legalese," *Northern Ireland Legal Quarterly* 68, no. 1 (2017): 90.

44. Stephenson, "Harry Potter Language."

45. Thomas W. LeBlanc, Ashley Hesson, Andrew Williams, Chris Feudtner, Margaret Holmes-Rovner, Lillie D. Williamson, and Peter A. Ubel, "Patient Understanding of Medical Jargon: A Survey Study of U.S. Medical Students," *Patient Education and Counseling* 95, no. 2 (2014): 238–42.

46. N. Ali, A. A. Khan, M. Akunjee, and F. Ahfat, "Using Common Ophthalmologic Jargon in Correspondence Can Lead to Miscommunication," *British Journal of General Practice*, December 2006, 968–69.

CHAPTER 5

1. Shellie Nelson, "Sen. Grassley Gets Attention for Tweeting 'u kno what,'" *WQAD News*, November 7, 2014, https://wqad.com/2014/11/07/sen-grassley -gets-attention-for-tweeting-u-kno-what.

2. A. Goldman, "What's Up with Chuck Grassley's Twitter Feed?" *Digg*, April 16, 2015, https://digg.com/2015/chuck-grassley-twitter.

3. Andrew Kaczynski, "Senator Grassley Explains His Odd Dairy Queen Tweet," *BuzzFeed*, November 7, 2014, https://www.buzzfeednews.com/article/ andrewkaczynski/senator-grassley-explains-his-odd-dairy-queen-tweet.

4. Michael Tayler and Jane Ogden, "Doctors' Use of Euphemisms and Their Impact on Patients' Beliefs about Health: An Experimental Study of Heart Failure," *Patient Education and Counseling* 57, no. 3 (2005): 321–26.

5. Deborah Rawlings, Jennifer J. Tieman, Christine Sanderson, Deborah Parker, and Lauren Miller-Lewis, "Never Say Die: Death Euphemisms, Misunderstandings and Their Implications for Practice," *International Journal of Palliative Nursing* 23, no. 7 (2017): 324–30.

6. William J. Astore, "All the Euphemisms We Use for 'War,'" *The Nation*, April 15, 2016.

7. W. Donald Smith, "Beyond the 'Bridge on the River Kwai': Labor Mobilization in the Greater East Asia Co-Prosperity Sphere," *International Labor and Working-Class History* 58 (2000): 219–38.

8. Morton Ann Gernsbacher, Adam R. Raimond, M. Theresa Balinghasay, and Jilana S. Boston, "'Special Needs' Is an Ineffective Euphemism," *Cognitive Research: Principles and Implications* 1, no. 1 (2016): 1–13.

9. Jeffrey K. Salkin, *Righteous Gentiles in the Hebrew Bible: Ancient Role Models for Sacred Relationships* (Woodstock, VT: Jewish Lights Publishing, 2008).

10. Alex Lubertozzi, "Introduction," in *Oliver Twist* (Oak Park, IL: Top Five Books, 2019).

11. Marvin Rosenberg, *The Adventures of a Shakespeare Scholar: To Discover Shakespeare's Art* (Newark: University of Delaware Press, 1997).

12. Daniel Gross, "That's What She Said: The Rise and Fall of the 2000s' Best Bad Joke," *The Atlantic*, January 24, 2014.

13. Juliane House, Gabriele Kasper, and Steven Ross, "Misunderstanding Talk," in *Misunderstanding in Social Life: Discourse Approaches to Problematic Talk*, ed. Juliane House, Gabriele Kasper, and Steven Ross (Harlow: Pearson Education, 2003).

14. Carol Reeves, *The Language of Science* (London: Routledge, 2005).

15. Keith S. Taber, "When the Analogy Breaks Down: Modelling the Atom on the Solar System," *Physics Education* 36, no. 3 (2001): 222–26.

16. George Lakoff and Mark Johnson, *Metaphors We Live By* (Chicago: University of Chicago Press, 1980).

17. Michael Reddy, "The Conduit Metaphor: A Case of Frame Conflict in Our Language about Language," in *Metaphor and Thought*, ed. Andrew Ortony (Cambridge: Cambridge University Press, 1979), 284–324.

18. Keith S. Taber, "Mind Your Language: Metaphor Can Be a Double-Edged Sword," *Physics Education* 40, no. 1 (2005): 11–12.

19. Andreas Musolff, "Metaphors: Sources for Intercultural Misunderstanding?" *International Journal of Language and Culture* 1, no. 1 (2014): 42–59.

20. Bernie Becker and Aaron Lorenzo, "Tax Writers See Peril in Trump's Obamacare Persistence," *Politico*, August 7, 2017, https://www.politico.com/story/2017/08/07/trump-obamacare-congress-tax-reform-241340.

21. Miles Klee, "Senator Orrin Hatch and the Origin of 'Shooting One's Wad,'" *Mel Magazine*, August 2017, https://melmagazine.com/en-us/story/senator-orrin-hatch-and-the-origin-of-shooting-ones-wad.

22. Gareth Carrol, Jeannette Littlemore, and Margaret Gillon Dowens, "Of False Friends and Familiar Foes: Comparing Native and Non-native Understanding of Figurative Phrases," *Lingua* 204 (2018): 21–44.

23. Matt Soniak, "How Did the Term 'Open a Can of Worms' Originate?" *Mental Floss*, June 28, 2012, https://www.mentalfloss.com/article/31039/how-did-term-open-can-worms-originate.

24. Lakoff and Johnson, *Metaphors We Live By*.

25. Raymond W. Gibbs Jr. and Jennifer E. O'Brien, "Idioms and Mental Imagery: The Metaphorical Motivation for Idiomatic Meaning," *Cognition* 36, no. 1 (1990): 35–68.

26. Robert Allen, *How to Write Better English* (London: Penguin Books, 2005).

27. Steven Pinker, *The Blank Slate: The Modern Denial of Human Nature* (New York: Penguin Books, 2003).

28. Jayne Tsuchiyama, "The Term 'Oriental' Is Outdated, but Is It Racist?" *Los Angeles Times*, June 1, 2016.

29. Henry Ansgar Kelly, "Rule of 'Thumb' and the Folklaw of the Husband's Stick," *Journal of Legal Education* 44 (1994): 341–65.

30. Patricia T. O'Conner and Stewart Kellerman, *Origins of the Specious: Myths and Misconceptions of the English Language* (New York: Random House, 2009).

31. Mara L. Grayson, *Teaching Racial Literacy: Reflective Practices for Critical Writing* (Lanham, MD: Rowman & Littlefield, 2018).

32. Kate Burridge and Tonya N. Stebbins, *For the Love of Language: An Introduction to Linguistics* (Cambridge: Cambridge University Press, 2016), 78.

33. Joe Satran, "Vintage Slang Terms for Being Drunk Are Hilarious a Century Later," *HuffPost*, December 7, 2017, https://www.huffpost.com/entry/vintage-slang-terms-drunk_n_4268480.

34. Wilson Andrews and Josh Katz, "Language Quiz: Are You on Fleek?" *New York Times*, February 22, 2015.

35. Lynne Kelly, James A. Keaten, Bonnie Becker, Jodi Cole, Lea Littleford, and Barrett Rothe, "'It's the American Lifestyle!': An Investigation of Text Messaging by College Students," *Qualitative Research Reports in Communication* 13, no. 1 (2012): 1–9.

36. Lynne Kelly and Aimee E. Miller-Ott, "Perceived Miscommunication in Friends' and Romantic Partners' Texted Conversations," *Southern Communication Journal* 83, no. 4 (2018): 270.

37. Kelly and Miller-Ott, "Perceived Miscommunication in Friends' and Romantic Partners' Texted Conversations."

38. Jonathan Ohadi, Brandon Brown, Leora Trub, and Lisa Rosenthal, "I Just Text to Say I Love You: Partner Similarity in Texting and Relationship Satisfaction,"*Computers in Human Behavior* 78 (2018): 126–32.

39. Reader's Digest, *Laughter Totally Is the Best Medicine* (White Plains, NY: Trusted Media Brands, Inc., 2019).

40. John Rentoul, "The Top 10: Commonly Confused Abbreviations," *The Independent*, February 14, 2020.

41. Michelle A. McSweeney, *The Pragmatics of Text Messaging: Making Meaning in Messages* (New York: Routledge, 2018).

42. Jillian Madison, *Damn You, Autcorrect! Awesomely Embarrassing Text Messages You Didn't Mean to Send* (New York: Hyperion, 2011).

43. Jerry Davich, "Text Messages Can Easily Get Lost in Translation," *Chicago Tribune*, June 29, 2017.

44. *Daily Mail*, "Jail for Knifeman Who Stabbed Friend to Death in Row Over Text Message," April 1, 2011.

CHAPTER 6

1. Stephen Porter, Leanne ten Brinke, and Chantal Gustaw, "Dangerous Decisions: The Impact of First Impressions of Trustworthiness on the Evaluation of Legal Evidence and Defendant Culpability," *Psychology, Crime, and Law* 16, no. 6 (2010): 477–91.

2. Martin Vestergaard, Mickey T. Kongerslev, Marianne S. Thomsen, Birgit Bork Mathiesen, Catherine J. Harmer, Erik Simonsen, and Kamilla W. Miskowiak, "Women with Borderline Personality Disorder Show Reduced Identification of Emotional Facial Expressions and a Heightened Negativity Bias," *Journal of Personality Disorders* 34, no. 5 (2020): 677–98.

3. Roy Azoulay, Uri Berger, Hadar Keshet, Paula M. Niedenthal, and Eva Gilboa-Schechtman, "Social Anxiety and the Interpretation of Morphed Facial Expressions Following Exclusion and Inclusion," *Journal of Behavior Therapy and Experimental Psychiatry* 66 (2020): 101–11.

4. Rishikesh V. Behere, "Facial Emotion Recognition Deficits: The New Face of Schizophrenia," *Indian Journal of Psychiatry* 57, no. 3 (2015): 229.

5. Andrew J. Calder, Jill Keane, Tom Manly, Reiner Sprengelmeyer, Sophie Scott, Ian Nimmo-Smith, and Andrew W. Young, "Facial Expression Recognition across the Adult Life Span," *Neuropsychologia* 41, no. 2 (2003): 195–202.

6. Madeline B. Harms, Alex Martin, and Gregory L. Wallace, "Facial Emotion Recognition in Autism Spectrum Disorders: A Review of Behavioral and Neuroimaging Studies," *Neuropsychology Review* 20, no. 3 (2010): 290–322.

7. Jessica Bennett, "I'm Not Mad. That's Just My RBF," *New York Times,* August 1, 2015.

8. Heidi Grant Halvorson, *No One Understands You and What to Do about It* (Boston: Harvard Business Review Press, 2015).

9. Dana Berkowitz, *Botox Nation: Changing the Face of America* (New York: New York University Press, 2017).

10. Lisa Feldman Barrett, Ralph Adolphs, Stacy Marsella, Aleix M. Martinez, and Seth D. Pollak, "Emotional Expressions Reconsidered: Challenges to Inferring Emotion from Human Facial Movements," *Psychological Science in the Public Interest* 20, no. 1 (2019): 1–68.

11. Tony Bravo, "How to 'Smize' Like You Mean It," *San Francisco Chronicle,* May 2, 2020.

12. Emese Nagy, "Is Newborn Smiling Really Just a Reflex? Research Is Challenging the Textbooks," *The Conversation,* October 30, 2018, https://theconversation.com/is-newborn-smiling-really-just-a-reflex-research-is-challenging-the-textbooks-105220.

13. Paul Ekman, Richard J. Davidson, and Wallace V. Friesen, "The Duchenne Smile: Emotional Expression and Brain Physiology: II," *Journal of Personality and Social Psychology* 58, no. 2 (1990): 342–53.

14. Eva G. Krumhuber and Antony S. R. Manstead, "Can Duchenne Smiles Be Feigned? New Evidence on Felt and False Smiles," *Emotion* 9, no. 6 (2009): 807–20.

15. Peter Hurley, *The Headshot: The Secrets to Creating Amazing Headshot Portraits* (San Francisco: New Riders Press, 2016).

16. Simon Baron-Cohen, Therese Jolliffe, Catherine Mortimore, and Mary Robertson, "Another Advanced Test of Theory of Mind: Evidence from Very High Functioning Adults with Autism or Asperger Syndrome," *Journal of Child Psychology and Psychiatry* 38, no. 7 (1997): 813–22.

17. Simon Baron-Cohen, Daniel C. Bowen, Rosemary J. Holt, Carrie Allison, Bonnie Auyeung, Michael V. Lombardo, Paula Smith, and Meng-Chuan Lai, "The 'Reading the Mind in the Eyes' Test: Complete Absence of Typical Sex Difference in ~400 Men and Women with Autism," *PLoS One* 10, no. 8 (2015): e0136521.

18. Joshua K. Hartshorne and Laura T. Germine, "When Does Cognitive Functioning Peak? The Asynchronous Rise and Fall of Different Cognitive Abilities across the Life Span," *Psychological Science* 26, no. 4 (2015): 433–43.

19. David Dodell-Feder, Kerry J. Ressler, and Laura T. Germine, "Social Cognition or Social Class and Culture? On the Interpretation of Differences in Social Cognitive Performance," *Psychological Medicine* 50, no. 1 (2020): 133–45.

20. Stella Marie Hombach, "From behind the Coronavirus Mask, an Unseen Smile Can Still Be Heard," *Scientific American*, June 1, 2020, https://www.scientificamerican.com/article/from-behind-the-coronavirus-mask-an-unseen-smile-can-still-be-heard.

21. Claus-Christian Carbon, "Wearing Face Masks Strongly Confuses Counterparts in Reading Emotions," *Frontiers in Psychology*, September 25, 2020, https://www.frontiersin.org/articles/10.3389/fpsyg.2020.566886/full.

22. David Boyle, *V for Victory: The Wireless Campaign That Defeated the Nazis* (London: The Real Press, 2016).

23. M. Paul Holsinger, ed., *War and American Popular Culture: A Historical Encyclopedia* (Westport, CT: Greenwood Press, 1999).

24. Stephanie Burnett, "Have You Ever Wondered Why East Asians Spontaneously Make V-Signs in Photos?" *Time*, August 3, 2014, https://time.com/2980357/asia-photos-peace-sign-v-janet-lynn-konica-jun-inoue.

25. Gayle Cotton, *Say Anything to Anyone, Anywhere: 5 Keys to Successful Cross-Cultural Communication* (Hoboken, NJ: John Wiley & Sons, 2013).

26. Cheryl Hamilton, *Communicating for Results: A Guide for Business and the Professions*, 10th ed. (Boston: Wadsworth Cengage, 2014).

27. Daniel Alacón, "How Do You Define a Gang Member?" *New York Times*, May 27, 2015.

28. Zoe Mintz, "Dontadrian Bruce, 15-Year-Old Student, Suspended after School Says Hand Gesture Represented 'Vice Lord' Gang Sign," *International Business Times*, March 12, 2014.

29. Rheanna Murray, "School Defends Principal in Controversial 'Gang Sign' Photo," *ABC News*, November 20, 2014.

30. Kia Makarechi, "'Pointergate' Is the Most Pathetic News Story of the Week," *Vanity Fair*, November 7, 2014.

31. Alexander Petri, "The Best Response to Pointergate, the Worst News Story of 2014?" *Washington Post*, November 14, 2014.

32. *New York Times*, "Virginia Tourism Agency Removes Gang Gesture from Ad," August 19, 2007.

33. Alex Halperin, "Did Urban Outfitters Mean to Put a Gang Sign on a Shirt?" *Salon*, July 18, 2013.

34. Nara Schoenberg, "OK Sign Is under Siege: How the Squeaky-Clean Hand Gesture Was Twisted by Trolls and Acquired Racist Undertones," *Chicago Tribune*, May 30, 2019.

35. Michael Levenson, "Military Says Hand Gestures at Game Were Not White Power Signs," *New York Times*, December 20, 2019.

36. Ben Smith, "I'll Take 'Hand Gestures of QAnon' for $1,000," *New York Times*, May 17, 2021, B1.

37. Associated Press, "Deaf Man Attacked because of Sign Language Mixup," March 4, 1995.

38. Philip Caulfield, "Group of Deaf, Mute Friends Stabbed at Bar after Thug Mistakes Sign Language for Gang Signs," *Daily Mail*, May 1, 2011.

39. Samantha Grossman, "Deaf Man Stabbed after Sign Language Mistaken for Gang Signs," *Time*, January 15, 2013.

40. Katy Steinmetz, "Oxford's 2015 Word of the Year Is This Emoji," *Time*, November 16, 2015.

41. Anne Quito, "Why We Can't Stop Using the 'Face with Tears of Joy' Emoji," *Quartz*, October 18, 2019, https://qz.com/1726756/the-psychology -behind-the-most-popular-emoji.

42. Lucia Peters, "Those Emoji Don't Mean What You Think They Do," *Bustle*, April 12, 2016.

43. Eric Goldman, "Emojis and the Law," *Washington Law Review* 93 (2018): 1227–91.

44. Vyvyan Evans, *The Emoji Code: The Linguistics behind Smiley Faces and Scaredy Cats* (New York: Picador, 2017).

45. Qiyu Bai, Qi Dan, Zhe Mu, and Maokun Yang, "A Systematic Review of Emoji: Current Research and Future Perspectives," *Frontiers in Psychology* 10 (2019): 2221.

46. Facebook Community Standards, Part III, Section 15, "Sexual Solicitation," August 2019, https://www.facebook.com/communitystandards/sexual _solicitation.

47. Hannah Frishberg, "'Sexual' Use of Eggplant and Peach Emojis Banned on Facebook, Instagram," *New York Post*, October 20, 2019.

48. Hannah Jean Miller, Jacob Thebault-Spieker, Shuo Chang, Isaac Johnson, Loren Terveen, and Brent Hecht, "'Blissfully Happy' or 'Ready to Fight': Varying Interpretations of Emoji," in *Proceedings of the Tenth International AAAI Conference on Web and Social Media*, ed. Markus Strohmaier and Krishna P. Gummadi (Palo Alto CA: AAAI Press, 2016), 259–68.

49. Hannah Miller, Daniel Kluver, Jacob Thebault-Spieker, Loren Terveen, and Brent Hecht, "Understanding Emoji Ambiguity in Context: The Role of

Text in Emoji-Related Miscommunication," in *Proceedings of the Eleventh International AAAI Conference on Web and Social Media*, ed. Derek Ruths (Palo Alto, CA: AAAI Press, 2017), 152–61.

50. Benjamin Weiser, "At Silk Road Trial, Lawyers Fight to Include Evidence They Call Vital: Emoji," *New York Times*, January 29, 2015, A22.

51. Matt Haber, "Should Grown Men Use Emoji?" *New York Times*, April 3, 2015, D27.

52. Gretchen McCulloch, *Because Internet: Understanding the New Rules of Language* (New York: Riverhead Books, 2019).

53. Jason Turbow, *The Baseball Codes: Beanballs, Sign Stealing, and Bench-Clearing Brawls: The Unwritten Rules of America's Pastime* (New York: Pantheon, 2010).

54. Michael S. Schmidt, "Phillies Are Accused of Stealing Signs Illegally," *New York Times*, May 12, 2010.

55. Michael S. Schmidt, "Boston Red Sox Used Apple Watches to Steal Signs against Yankees," *New York Times*, September 5, 2017.

56. Neil Vigdor, "The Houston Astros' Cheating Scandal: Sign-Stealing, Buzzer Intrigue and Tainted Pennants," *New York Times*, July 16, 2020.

57. USA Today Network, "What Do All Those Hand Signs Mean? Inside the Hidden Language of Baseball and Softball," *Wausau Daily Herald*, April 24, 2018.

58. Tim Rohan, "Mets Revamp Sign System after Daniel Murphy's Departure," *New York Times*, March 6, 2016.

59. Jack Curry, "Mets Follow the Wrong Sign to Another Defeat," *New York Times*, June 14, 1991, B11, B13.

60. Sam Farmer, "Ask Farmer: How Many Plays in an NFL Playbook, and How Many Do Teams Run in a Game?" *Los Angeles Times*, January 3, 2016.

61. Angelique Chengelis, "Big Picture: Play Cards Might Look Silly but Are 'Very Important' in Michigan Offense," *Detroit News*, September 19, 2019.

62. Jesse Newell, "How Miscommunication Hurt KU Football's Offense in Two Underwhelming Games," *Kansas City Star*, September 9, 2019.

63. Alex Ross, "Searching for Silence: John Cage's Art of Noise," *The New Yorker*, October 4, 2010.

64. Timothy D. Wilson, David A. Reinhard, Erin C. Westgate, Daniel T. Gilbert, Nicole Ellerbeck, Cheryl Hahn, Casey L. Brown, and Adi Shaked, "Just Think: The Challenges of the Disengaged Mind," *Science* 345, no. 6192 (2014): 75–77.

65. Takie Sugiyama Lebra, "The Cultural Significance of Silence in Japanese Communication," *Multilingua* 6, no. 4 (1987): 343–58.

66. Kurzon, Dennis, "Towards a Typology of Silence," *Journal of Pragmatics* 39, no. 10 (2007): 1673–88.

67. Linda B. Gambrell, "The Occurrence of Think-Time during Reading Comprehension Instruction," *Journal of Educational Research* 77, no. 2 (1983): 77–80.

68. Barbara A. Wasik and Annemarie H. Hindman, "Why Wait? The Importance of Wait Time in Developing Young Students' Language and Vocabulary Skills," *The Reading Teacher* 72, no. 3 (2018): 369–78.

69. Herbert H. Clark and Jean E. Fox Tree, "Using 'uh' and 'um' in Spontaneous Speaking," *Cognition* 84, no. 1 (2002): 73–111.

70. Shirley Näslund, "Tacit Tango: The Social Framework of Screen-Focused Silence in Institutional Telephone Calls," *Journal of Pragmatics* 91 (2016): 60–79.

71. Milena Popova, *Sexual Consent* (Cambridge, MA: MIT Press, 2019).

CHAPTER 7

1. N. G. N. Kelsey, *Games, Rhymes, and Wordplay of London Children* (London: Palgrave Macmillan, 2019), 131.

2. Tomasz P. Krzeszowski, "The Bible Translation Imbroglio," in *Cultural Conceptualization in Translation and Language Applications*, ed. Barbara Lewandwoska-Tomaszczyk (Cham: Springer Nature Switzerland, 2020), 15–33.

3. Arthur Asseraf, *Electronic News in Colonial Algeria* (Oxford: Oxford University Press, 2019).

4. Eleanor Ochs, "Misunderstanding Children," in *"Miscommunication" and Problematic Talk*, ed. Nikolas Coupland, Howard Giles, and John M. Wiemann (Newbury Park, CA: Sage Publications, 1991), 44–60.

5. Eve V. Clark, *First Language Acquisition*, 2nd ed. (Cambridge: Cambridge University Press, 2009).

6. Mary K. Fagan, "Toddlers' Persistence When Communication Fails: Response Motivation and Goal Substitution," *First Language* 28, no. 1 (2008): 55–69.

7. Ora Aviezer, "Bedtime Talk of Three-Year-Olds: Collaborative Repair of Miscommunication," *First Language* 23, no. 1 (2003): 117–39.

8. Sarah A. Bacso and Elizabeth S. Nilsen, "What's That You're Saying? Children with Better Executive Functioning Produce and Repair

Communication More Effectively," *Journal of Cognition and Development* 18, no. 4 (2017): 441–64.

9. Elisabeth S. Pasquini, Kathleen H. Corriveau, Melissa Koenig, and Paul L. Harris, "Preschoolers Monitor the Relative Accuracy of Informants," *Developmental Psychology* 43, no. 5 (2007): 1216–26.

10. George A. Miller, Eugene Galanter, and Karl H. Pribram, *Plans and the Structure of Behavior* (New York: Henry Holt, 1960), 153.

11. Thomas G. Bever, "The Cognitive Basis for Linguistic Structures," in *Cognition and the Development of Language*, ed. J. R. Hayes (New York: John Wiley & Sons, 1970), 1–61.

12. H. W. Fowler, *A Dictionary of Modern English Usage* (Oxford: Oxford University Press, 1965).

13. John Kimball, "Seven Principles of Surface Structure Parsing in Natural Language," *Cognition* 2 (1973): 12–47.

14. Charles A. Perfetti, Sylvia Beverly, Laura Bell, Kimberly Rodgers, and Robert Faux, "Comprehending Newspaper Headlines," *Journal of Memory and Language* 26, no. 6 (1987): 692–713.

15. Chiara Bucaria, "Lexical and Syntactic Ambiguity as a Source of Humor: The Case of Newspaper Headlines," *Humor* 17, no. 3 (2004): 279–310.

16. Ben Zimmer, "Crash Blossoms," *New York Times Magazine*, January 31, 2010, MM14.

17. Chris Elliott, "The Reader's Editor on . . . How Headlines Can Be More Easily Misunderstood Online," *The Guardian*, September 16, 2012.

18. Benjamin A. Levett, *Through the Customs Maze: A Popular Exposition and Analysis of the United States Customs Tariff Administrative Laws* (New York: Customs Maze Publishing, 1923).

19. *New York Times*, "Hyphen or Comma? The History of the Error in the Enrollment Tariff Bill," February 21, 1874.

20. Lynne Truss, *Eats, Shoots & Leaves: The Zero Tolerance Approach to Punctuation* (New York: Avery, 2003).

21. Mary Norris, *Between You & Me: Confessions of a Comma Queen* (New York: W. W. Norton, 2015), 93.

22. Daniel Victor, "Oxford Comma Dispute Is Settled as Maine Drivers Get $5 Million," *New York Times*, February 9, 2018.

23. Adam Freedman, "Clause and Effect," *New York Times*, December 16, 2007.

24. Cecelia Watson, *Semicolon: The Past, Present, and Future of a Misunderstood Mark* (New York: Ecco, 2019).

25. Livia Albeck-Ripka, "Missing Apostrophe in Facebook Post Lands a Man in Defamation Court," *New York Times*, October 11, 2021.

26. Richard P. Heitz, "The Speed-Accuracy Tradeoff: History, Physiology, Methodology, and Behavior," *Frontiers in Neuroscience* 8 (2014): 150.

27. Mark Seidenberg, *Language at the Speed of Sight: How We Read, Why So Many Can't, and What Can Be Done about It* (New York: Basic Books, 2017).

28. Keith Rayner, Elizabeth R. Schotter, Michael E. J. Masson, Mary C. Potter, and Rebecca Treiman, "So Much to Read, So Little Time: How Do We Read, and Can Speed Reading Help?" *Psychological Science in the Public Interest* 17, no. 1 (2016): 4–34.

29. Dina Acklin and Megan H. Papesh, "Modern Speed-Reading Apps Do Not Foster Reading Comprehension," *American Journal of Psychology* 130, no. 2 (2017): 183–99.

30. Seidenberg, *Language at the Speed of Sight*.

31. Steve Tauroza and Desmond Allison, "Speech Rates in British English," *Applied Linguistics* 11, no. 1 (1990): 90–105.

32. Gareth Walker, "The Phonetic Constitution of a Turn-Holding Practice: Rush-Throughs in English Talk-in-Interaction," in *Prosody in Interaction*, ed. Dagmar Barth-Weingarten, Elisabeth Reber, and Margret Selting (Amsterdam: John Benjamins, 2010), 51–72.

33. Raymond Pastore and Albert D. Ritzhaupt, "Using Time-Compression to Make Multimedia Learning More Efficient: Current Research and Practice," *TechTrends* 59, no. 2 (2015): 66–74.

34. Ray Pastore, "The Effects of Time-Compressed Instruction and Redundancy on Learning and Learners' Perceptions of Cognitive Load," *Computers and Education* 58, no. 1 (2012): 641–51.

35. Ray Pastore, "Time-Compressed Instruction: What Compression Speeds Do Learners Prefer?" *International Journal of Instructional Technology and Distance Learning* 12, no. 6 (2015): 3–20.

36. Henry L. Roediger and K. Andrew DeSoto, "Recognizing the Presidents: Was Alexander Hamilton President?" *Psychological Science* 27, no. 5 (2016): 644–50.

37. Aaron French, "The Mandela Effect and New Memory," *Correspondences: Journal for the Study of Esotericism* 6, no. 2 (2018): 201–33.

38. Fiona Broome, *The Mandela Effect—Major Memories, Book 1* (N.p.: New Forest Books, 2019).

39. Henry L. Roediger and Kathleen B. McDermott, "Creating False Memories: Remembering Words Not Presented in Lists," *Journal of Experimental Psychology: Learning, Memory, and Cognition* 21, no. 4 (1995): 803–14.

40. Frederic C. Bartlett, *Remembering: A Study in Experimental and Social Psychology* (Cambridge: Cambridge University Press, 1932/1997).

41. Nathaniel Hébert, "Sex IN THE City; Carrie Bradshaw and the Mandela Effect," *Medium*, May 23, 2018, https://medium.com/@nathanielhebert/sex-in-the-city-carrie-bradshaw-and-the-mandela-effect-bead3a66b7bf.

42. Mark Davies, "Expanding Horizons in Historical Linguistics with the 400-Million Word Corpus of Historical American English," *Corpora* 7, no. 2 (2012): 121–57.

43. Caitlin Aamodt, "Collective False Memories: What's behind the 'Mandela Effect'?" *Discover*, February 16, 2017, https://www.discovermagazine.com/mind/collective-false-memories-whats-behind-the-mandela-effect.

44. Roediger and DeSoto, "Recognizing the Presidents."

45. Eli Yokley, "Voters Want Gary Johnson, Jill Stein on the Debate Stage," *Morning Consult*, September 1, 2016, https://morningconsult.com/2016/09/01/voters-want-gary-johnson-jill-stein-debate-stage.

46. Jonah Bromwich, "'I Guess I'm Having an Aleppo Moment': Gary Johnson Can't Name a Single Foreign Leader," *New York Times*, September 28, 2016.

47. John Hendrickson, "Gary Johnson Is Ready to Talk about Aleppo," *Esquire*, February 23, 2018.

48. Ryan Teague Beckwith, "Read the Interview Where Gary Johnson Asked What Aleppo Is," *Time*, September 8, 2016.

49. Wilson L. Taylor, "'Cloze Procedure': A New Tool for Measuring Readability," *Journalism Bulletin* 30, no. 4 (1953): 415–33.

50. Richard A. Oppel Jr., "After Debate Gaffe, Perry Vows to Stay in Race," *New York Times*, November 10, 2011.

51. Gitit Kavé, Ariel Knafo, and Asaf Gilboa, "The Rise and Fall of Word Retrieval across the Lifespan," *Psychology and Aging* 25, no. 3 (2010): 719–24.

CHAPTER 8

1. Dacher Keltner, Lisa Capps, Ann M. Kring, Randall C. Young, and Erin A. Heerey, "Just Teasing: A Conceptual Analysis and Empirical Review," *Psychological Bulletin* 127, no. 2 (2001): 229–48.

2. Robin M. Kowalski, "'I Was Only Kidding!': Victims' and Perpetrators' Perceptions of Teasing," *Personality and Social Psychology Bulletin* 26, no. 2 (2000): 231–41.

3. Dacher Keltner, Randall C. Young, Erin A. Heerey, Carmen Oemig, and Natalie D. Monarch, "Teasing in Hierarchical and Intimate Relations," *Journal of Personality and Social Psychology* 75, no. 5 (1998): 1231–47.

4. Carol Bishop Mills, "Child's Play or Risky Business? The Development of Teasing Functions and Relational Implications in School-Aged Children," *Journal of Social and Personal Relationships* 35, no. 3 (2018): 287–306.

5. Justin Kruger, Cameron L. Gordon, and Jeff Kuban, "Intentions in Teasing: When 'Just Kidding' Just Isn't Good Enough," *Journal of Personality and Social Psychology* 90, no. 3 (2006): 412–25.

6. Michael Haugh, "'Just Kidding': Teasing and Claims to Non-Serious Intent," *Journal of Pragmatics* 95 (2016): 120–36.

7. Nilupama Wijewardena, Ramanie Samaratunge, Charmine Härtel, and Andrea Kirk-Brown, "Why Did the Emu Cross the Road? Exploring Employees' Perception and Expectations of Humor in the Australian Workplace," *Australian Journal of Management* 41, no. 3 (2016): 563–84.

8. Glen Gorman and Christian H. Jordan, "'I Know You're Kidding': Relationship Closeness Enhances Positive Perceptions of Teasing," *Personal Relationships* 22, no. 2 (2015): 173–87.

9. Richard M. Roberts and Roger J. Kreuz, "Why Do People Use Figurative Language?" *Psychological Science* 5, no. 3 (1994): 159–63.

10. Derrick B. Taylor, "Professor Fired after Joking That Iran Should Pick U.S. Sites to Bomb," *New York Times*, January 11, 2020.

11. Niraj Chokshi, "That Wasn't Mark Twain: How a Misquotation Is Born," *New York Times*, April 26, 2017.

12. Noam Cohen, "Spinning a Web of Lies at Digital Speed," *New York Times*, October 12, 2008, B3.

13. Thomas L. Friedman, "Too Good to Check," *New York Times*, November 17, 2010, A33.

14. Andrew Rosenthal, "Dukakis Criticizes Bush Tax Cut Plan," *New York Times*, October 25, 1988, A26.

15. Nicholas DiFonzo and Prashant Bordia, "Rumor, Gossip, and Urban Legends," *Diogenes* 213 (2007): 19–35.

16. C. J. Walker and D. Struzyk, "Evidence for a Social Conduct Moderating Function of Common Gossip," paper presented to the International Society for the Study of Close Relationships, Saratoga Springs, NY, 1998.

17. Aaron Ben-Ze'ev, "The Vindication of Gossip," in *Good Gossip*, ed. Robert F. Goodman and Aaron Ben-Ze'ev (Lawrence: University Press of Kansas, 1994), 11–24.

18. Eric K. Foster, "Research on Gossip: Taxonomy, Methods, and Future Directions," *Review of General Psychology* 8, no. 2 (2004): 78–99.

19. L. M. Montgomery, *Chronicles of Avonlea* (Toronto: McClelland & Stewart, 1912).

20. Gordon Allport and Leo Postman, *The Psychology of Rumor* (New York: Henry Holt, 1947).

21. Anthony J. Bocchino and David A. Sonenshein, *A Practical Guide to Federal Evidence: Objections, Responses, Rules, and Practice Commentary*, eighth edition (Louisville, CO: National Institute for Trial Advocacy, 2006).

22. Luiza Newlin-Łukowicz, "TH-Stopping in New York City: Substrate Effect Turned Ethnic Marker?" *University of Pennsylvania Working Papers in Linguistics* 19, no. 2 (2013): 150–60.

23. Luanne Von Schneidemesser, "Soda or Pop?" *Journal of English Linguistics* 24, no. 4 (1996): 270–87.

24. Alan McConchie, "Pop vs Soda," http://www.popvssoda.com.

25. Richard Gardiner, "The Civil War Origin of Coca-Cola in Columbus, Georgia," *Muscogee Genealogical Society* 2012: 21–24.

26. Edwin L. Battistella, "The Syntax of the Double Modal Construction," *Linguistica Atlantica* 17 (1995): 19–44.

27. Margaret Mishoe and Michael Montgomery, "The Pragmatics of Multiple Modal Variation in North and South Carolina," *American Speech* 69, no. 1 (1994): 3–29.

28. Donald H. Naftulin, John E. Ware, and Frank A. Donnelly, "The Doctor Fox Lecture: A Paradigm of Educational Seduction," *Journal of Medical Education* 48, no. 7 (1973): 630–35.

29. John E. Ware and Reed G. Williams, "The Dr. Fox Effect: A Study of Lecturer Effectiveness and Ratings of Instruction," *Journal of Medical Education* 50, no. 2 (1975): 149–56.

30. Eyal Peer and Elisha Babad, "The Doctor Fox Research (1973) Rerevisited: 'Educational Seduction' Ruled Out," *Journal of Educational Psychology* 106, no. 1 (2014): 36–45.

31. Richard E. Petty and John T. Cacioppo, "The Elaboration Likelihood Model of Persuasion," in *Advances in Social Psychology*, no. 19, ed. Leonard Berkowitz (New York: Springer, 1986), 123–205.

32. David Mikkelson, "Did Barack Obama Say He Had Visited 57 (Islamic) States?" *Snopes*, June 19, 2008, https://www.snopes.com/fact-check/57-states.

33. Herbert H. Clark and Edward F. Schaefer, "Concealing One's Meaning from Overhearers," *Journal of Memory and Language* 26, no. 2 (1987): 209–25.

34. John H. Fleming, John M. Darley, James L. Hilton, and Brian A. Kojetin, "Multiple Audience Problem: A Strategic Communication Perspective on Social Perception," *Journal of Personality and Social Psychology* 58, no. 4 (1990): 593–609.

35. Roger Kreuz, *Irony and Sarcasm* (Cambridge, MA: MIT Press, 2020).

36. Sheldon M. Stern, *Averting "The Final Failure": John F. Kennedy and the Secret Cuban Missile Crisis Meetings* (Stanford, CA: Stanford University Press, 2003).

37. H. P. Grice, "Logic and Conversation," in *Syntax and Semantics, Volume 3: Speech Acts*, ed. P. Cole and J. L. Morgan (New York: Academic Press, 1975), 41–58.

38. Alex I. Thompson, "Wrangling Tips: Entrepreneurial Manipulation in Fast-Food Delivery," *Journal of Contemporary Ethnography* 44, no. 6 (2015): 737–65.

39. Orin S Kerr, "The Decline of the Socratic Method at Harvard," *Nebraska Law Review* 78 (1999): 113–34.

40. Jill Anderson, "Misreading Like a Lawyer: Cognitive Bias in Statutory Interpretation," *Harvard Law Review* 127 no. 6 (2014): 1521–92.

41. Harold Bloom, *A Map of Misreading* (Oxford: Oxford University Press, 1975).

CHAPTER 9

1. Alice E. Marwick and danah boyd, "I Tweet Honestly, I Tweet Passionately: Twitter Users, Context Collapse, and the Imagined Audience," *New Media & Society* 13, no. 1 (2011): 114–33.

2. Teresa Gil-Lopez, Cuihua Shen, Grace A. Benefield, Nicholas A. Palomares, Michal Kosinski, and David Stillwell, "One Size Fits All: Context Collapse, Self-Presentation Strategies and Language Styles on Facebook," *Journal of Computer-Mediated Communication* 23, no. 3 (2018): 127–45.

3. John Koblin, "After Racist Tweet, Roseanne Barr's Show Is Canceled by ABC," *New York Times*, May 29, 2018.

4. Jacey Fortin, "Kevin Hart Steps Down as Oscars Host after Criticism Over Homophobic Tweets," *New York Times*, December 6, 2018.

5. Ella Torres, "Kathy Zhu, Miss Michigan 2019, Stripped of Her Title Over 'Offensive' Social Media Posts," *ABC News*, July 20, 2019, https://abcnews.go.com/US/kathy-zhu-miss-michigan-2019-stripped-title-offensive/story?id=64459158.

6. Abby Ohlheiser, "Ken Bone Was a 'Hero'. Now Ken Bone Is 'Bad'. It Was His Destiny as a Human Meme," *Washington Post*, October 14, 2016.

7. Jon Ronson, "How One Stupid Tweet Blew Up Justine Sacco's Life," *New York Times*, February 12, 2015.

8. Sam Biddle, "Justine Sacco Is Good at Her Job, and How I Came to Peace with Her," *Gawker*, December 20, 2014, https://gawker.com/justine -sacco-is-good-at-her-job-and-how-i-came-to-pea-1653022326.

9. Anna Wiener, *Uncanny Valley: A Memoir* (New York: Farrar, Straus and Giroux, 2020).

10. William Davis, "Really, This Jargon Is Getting Out of Hand," *New York Times*, March 25, 1973, 3.

11. Janet Shilling, "I'll Fight Jargon with Jargon," *New York Times*, August 22, 1987, sec. 1, 27.

12. Sana Siwolop, "Business; Counting Coinages in Jargon," *New York Times*, August 1, 1999, sec. 3, 6.

13. Marilyn Katzman, "Baffled by Office Buzzwords," *New York Times*, July 4, 2015.

14. Emma Green, "The Origins of Office Speak: What Corporate Buzzwords Reveal about the History of Work," *The Atlantic*, April 24, 2014.

15. Molly Young, "Garbage Language: Why Do Corporations Speak the Way They Do?" *New York*, February 17, 2020, https://www.thecut .com/2020/02/spread-of-corporate-speak.html.

16. Mark Morgioni, "Defending 'Garbage Language,' the Silly Corporate Terminology That Seriously Works," *Slate*, February 20, 2020, https://slate .com/human-interest/2020/02/garbage-language-business-speak-defense.html.

17. Howard Gardner, "Getting Acquainted with Jean Piaget," *New York Times*, January 3, 1979, C1.

18. Adam Kirsch, "Should an Author's Intentions Matter?" *New York Times*, March 10, 2015, 31.

19. Roland Barthes, *Image-Music-Text*, trans. Stephen Heath (New York: Hill and Wang, 1977).

20. Elaheh Fadaee, "Symbols, Metaphors and Similes in Literature: A Case Study of 'Animal Farm,'" *Journal of English and Literature* 2, no. 2 (2011): 19–27.

21. Ashley Marshall, "Daniel Defoe as Satirist," *Huntington Library Quarterly* 70, no. 4 (2007): 553–76.

22. Kate Loveman, *Reading Fictions, 1660–1740: Deception in English Literary and Political Culture* (Aldershot: Ashgate, 2008).

23. J. C. Carlier and C. T. Watts, "Roland Barthes's Resurrection of the Author and Redemption of Biography," *Cambridge Quarterly* 29, no. 4 (2000): 386–93.

24. Steve Sohmer, *Reading Shakespeare's Mind* (Manchester: Manchester University Press, 2017).

25. Mark Polizzotti, *Sympathy for the Traitor: A Translation Manifesto* (Cambridge, MA: MIT Press, 2018).

26. David Strauss, *The Planet Mars: A History of Observation and Discovery* (Tucson: University of Arizona Press, 1996).

27. Govert Schilling, *Atlas of Astronomical Discoveries* (New York: Springer Science & Business Media, 2011).

28. K. Maria D. Lane, *Geographies of Mars: Seeing and Knowing the Red Planet* (Chicago: University of Chicago Press, 2011).

29. Dennis Prager, *The Rational Bible: Exodus, God, Slavery, and Freedom* (Washington, DC: Regnery Faith, 2018).

30. RT, "Interpreter of Khrushchev's 'We Will Bury You' Phrase Dies at 81," May 17, 2014, https://www.rt.com/news/159524-sukhodrev-interpreter -khrushchev-cold-war.

31. Nataly Kelly and Jost Zetzsche, *Found in Translation: How Language Shapes Our Lives and Transforms the World* (New York: Perigee, 2012).

32. William Taubman, *Khrushchev: The Man and His Era* (New York: Norton, 2003).

33. David Shahar and Mark G. L. Sayers, "Prominent Exostosis Projecting from the Occipital Squama More Substantial and Prevalent in Young Adult Than Older Age Groups," *Scientific Reports* 8, no. 1 (2018): 3354.

34. Zaria Gorvett, "How Modern Life Is Transforming the Human Skel- eton," *BBC Future*, June 13, 2019.

35. Isaac Stanley-Becker, "'Horns' Are Growing on Young People's Skulls. Phone Use Is to Blame, Research Suggests," *Washington Post*, June 25, 2019.

36. Hannah Sparks, "Young People Are Growing Horns from Cellphone Use: Study," *New York Post*, June 20, 2019.

37. Stanley-Becker, "'Horns' Are Growing on Young People's Skulls."

38. Denise Grady, "About the Idea That You're Growing Horns from Look- ing Down at Your Phone . . . ," *New York Times*, June 20, 2019.

39. Jamie Ducharme, "No, Teenagers Are Not Growing 'Skull Horns' because of Smartphones," *Time*, June 21, 2019, https://time.com/5611036/ teenagers-skull-horns.

40. Beth Mole, "Debunked: The Absurd Story about Smartphones Causing Kids to Sprout Horns," *Ars Technica*, June 21, 2019, https://arstechnica.com/

science/2019/06/debunked-the-absurd-story-about-smartphones-causing-kids
-to-sprout-horns.

41. Frances H. Rauscher, Gordon L. Shaw, and Catherine N. Ky, "Music and Spatial Task Performance," *Nature* 365, no. 6447 (1993): 611.

42. Kevin Sack, "Georgia's Governor Seeks Musical Start for Babies," *New York Times*, January 15, 1998.

43. Samuel A. Mehr, "Miscommunication of Science: Music Cognition Research in the Popular Press," *Frontiers in Psychology* 6 (2015): 988.

44. Andy Newman and Sarah Maslin Nir, "New York Today: Cucumber Time," *New York Times*, August 19, 2013.

45. Allan Bell, "Media (Mis)communication on the Science of Climate Change," *Public Understanding of Science 3*, no. 3 (2016): 259–73.

46. Miriam Shuchman and Michael S. Wilkes, "Medical Scientists and Health News Reporting: A Case of Miscommunication," *Annals of Internal Medicine* 126, no. 12 (1997): 976–82.

47. Ann Henderson-Sellers, "Climate Whispers: Media Communication about Climate Change," *Climatic Change* 40, no. 3-4 (1998): 421–56.

48. David F. Ransohoff and Richard M. Ransohoff, "Sensationalism in the Media: When Scientists and Journalists May Be Complicit Collaborators," *Effective Clinical Practice* 4, no. 4 (2001): 185–88.

49. Susan A. Speer, "Flirting: A Designedly Ambiguous Action?" *Research on Language and Social Interaction* 50, no. 2 (2017): 128–50.

50. David M. Buss, "The Evolution of Human Intrasexual Competition: Tactics of Mate Attraction," *Journal of Personality and Social Psychology* 54, no. 4 (1988): 616–28.

51. Michael R. Cunningham, "Reactions to Heterosexual Opening Gambits: Female Selectivity and Male Responsiveness," *Personality and Social Psychology Bulletin* 15, no. 1 (1989): 27–41.

52. Mark S. Carey, "Nonverbal Openings to Conversation," paper presented at the annual meeting of the Eastern Psychological Association, Philadelphia, 1974.

53. Debra G. Walsh and Jay Hewitt, "Giving Men the Come-On: Effect of Eye Contact and Smiling in a Bar Environment," *Perceptual and Motor Skills* 61, no. 3 (1985): 873–74.

54. Nicolas Guéguen, Jacques Fischer-Lokou, Liv Lefebvre, and Lubomir Lamy, "Women's Eye Contact and Men's Later Interest: Two Field Experiments," *Perceptual and Motor Skills* 106, no. 1 (2008): 63–66.

55. Jonathan D. Huber and Edward S. Herold, "Sexually Overt Approaches in Singles Bars," *Canadian Journal of Human Sexuality* 15, no. 3/4 (2006): 133–46.

56. Betty H. La France, "What Verbal and Nonverbal Communication Cues Lead to Sex: An Analysis of the Traditional Sexual Script," *Communication Quarterly* 58, no. 3 (2010): 297–318.

57. David Dryden Henningsen, "Flirting with Meaning: An Examination of Miscommunication in Flirting Interactions," *Sex Roles* 50, no. 7–8 (2004): 481–89.

58. Brandi N. Frisby, Megan R. Dillow, Shelbie Gaughan, and John Nordlund, "Flirtatious Communication: An Experimental Examination of Perceptions of Social-Sexual Communication Motivated by Evolutionary Forces," *Sex Roles* 64, no. 9–10 (2011): 682–94.

59. Janet Halley, "The Move to Affirmative Consent," *Signs: Journal of Women in Culture and Society* 42, no. 1 (2016): 257–79.

CHAPTER 10

1. Kristin Byron, "Carrying Too Heavy a Load? The Communication and Miscommunication of Emotion by Email," *Academy of Management Review* 33, no. 2 (2008): 309–27.

2. Sarah Schafer, "E-Mail's Impersonal Tone Easily Misunderstood: Conflicts Can Arise from Mistyped, Misperceived Messages," *Washington Post*, November 3, 2000.

3. Justin Kruger, Nicholas Epley, Jason Parker, and Zhi-Wen Ng, "Egocentrism Over E-Mail: Can We Communicate as Well as We Think?" *Journal of Personality and Social Psychology* 89, no. 6 (2005): 925–36.

4. Monica A. Riordan and Lauren A. Trichtinger, "Overconfidence at the Keyboard: Confidence and Accuracy in Interpreting Affect in E-Mail Exchanges," *Human Communication Research* 43, no. 1 (2017): 1–24.

5. Christoph Laubert and Jennifer Parlamis, "Are You Angry (Happy, Sad) or Aren't You? Emotion Detection Difficulty in Email Negotiation," *Group Decision and Negotiation* 28, no. 2 (2019): 377–413.

6. Byron, "Carrying too Heavy a Load?"

7. Byron, "Carrying too Heavy a Load?"

8. M. Mahdi Roghanizad and Vanessa K. Bohns, "Ask in Person: You're Less Persuasive Than You Think Over Email," *Journal of Experimental Social Psychology* 69 (2017): 223–26.

9. Roghanizad and Bohns, "Ask in Person," 223.

10. *Kitsap Sun*, "Road Rage: Man Shot When Trying to Apologize," December 29, 1997, https://products.kitsapsun.com/archive/1997/12-29/0020 _road_rage__man_shot_when_trying_t.html.

11. Kim Murphy, "Driven to Extremes in the Northwest," *Los Angeles Times*, January 11, 1998.

12. Reginald G. Smart, Robert E. Mann, and Gina Stoduto, "The Prevalence of Road Rage," *Canadian Journal of Public Health* 94, no. 4 (2003): 247–50.

13. Mark J. M. Sullman, "The Expression of Anger on the Road," *Safety Science* 72 (2015): 153–59.

14. Michael Fumento, "'Road Rage' versus Reality," *Atlantic Monthly* 282, no. 2 (August 1998): 12–17.

15. Joel Best and Mary M. Hutchinson, "The Gang Initiation Rite as a Motif in Contemporary Crime Discourse," *Justice Quarterly* 13, no. 3 (1996): 383–404.

16. Leslie Kendrick, "A Test for Criminally Instructional Speech," *Virginia Law Review* 91 (2005): 1973–2021.

17. John Tierney, "The Big City; A Hand Signal to Counteract Road Rage," *New York Times*, March 15, 1999, B1.

18. Tom Magliozzi and Ray Magliozzi, "The Horn and Lights Say It All in Basic Language of the Road," *Orlando Sentinel*, April 19, 2001.

19. Paul Bisceglio, "Why Don't Cars Have a 'Sorry' Signal?" *Pacific Standard*, February 3, 2014, https://psmag.com/environment/dont-cars-sorry -signal-73890.

20. Joshua Dressler, "Rethinking Heat of Passion: A Defense in Search of a Rationale," *Journal of Criminal Law and Criminology* 73 (1982): 421–70.

21. Randolph N. Jonakait, Harold Baer Jr., E. Stewart Jones Jr., and Edward J. Imwinkelried, *New York Evidentiary Foundations*, 2nd ed. (Charlottesville, VA: The Michie Company, 1998).

22. Craig Hemmens, Kathryn E. Scarborough, and Rolando V. Del Carmen, "Grave Doubts about 'Reasonable Doubt': Confusion in State and Federal Courts," *Journal of Criminal Justice* 25, no. 3 (1997): 231–54.

23. Irwin A. Horowitz, "Reasonable Doubt Instructions: Commonsense Justice and Standard of Proof," *Psychology, Public Policy, and Law* 3, no. 2–3 (1997): 285–302.

24. Taylor Jones, Jessica Rose Kalbfeld, Ryan Hancock, and Robin Clark, "Testifying while Black: An Experimental Study of Court Reporter Accuracy

in Transcription of African American English," *Language* 95, no. 2 (2019): e216–52.

25. John R. Rickford and Sharese King, "Language and Linguistics on Trial: Hearing Rachel Jeantel (and Other Vernacular Speakers) in the Courtroom and Beyond," *Language* 92, no. 4 (2016): 948–88.

26. Dana Chipkin, *Successful Freelance Court Reporting* (Albany, NY: Delmar West Legal Studies, 2001).

27. *New York Times*, "Answers from a Court Reporter," June 16, 2010.

28. Jones et al., "Testifying While Black."

29. Christopher Hibbert, *The Destruction of Lord Raglan: A Tragedy of the Crimean War, 1854–55* (London: Longmans, 1961).

30. Terry Brighton, *Hell Riders: The True Story of the Charge of the Light Brigade* (New York: Henry Holt, 2004).

31. Ezra J. Warner, *Generals in Gray: Lives of the Confederate Commanders* (Baton Rouge: Louisiana State University Press, 2006).

32. Mathew W. Lively, *Calamity at Chancellorsville: The Wounding and Death of Confederate General Stonewall Jackson* (El Dorado Hills, CA: Savas Beatie, 2013).

33. Rodney P. Carlisle, *Persian Gulf War* (New York: Facts on File, 2009).

34. David Zucchino and David S. Cloud, "U.S. Deaths in Drone Strike Due to Miscommunication, Report Says," *Los Angeles Times*, October 14, 2011.

35. Kelsey Munro, "How Safe Is Flying? Here's What the Statistics Say," *SBSNews*, July 31, 2018, https://www.sbs.com.au/news/how-safe-is-flying-here -s-what-the-statistics-say.

36. William Rankin, "MEDA Investigation Process," *Boeing Aero* 26, no. 2 (2007): 15–21.

37. J. Charles Alderson, "Air Safety, Language Assessment Policy, and Policy Implementation: The Case of Aviation English," *Annual Review of Applied Linguistics* 29 (2009): 168–87.

38. Barbara Clark, *Aviation English Research Project: Data Analysis Findings and Best Practice Recommendations* (Gatwick: Civil Aviation Authority, 2017).

39. John W. Howard III, "'Tower, Am I Cleared to Land?': Problematic Communication in Aviation Discourse," *Human Communication Research* 34, no. 3 (2008): 370–91.

40. Clark, *Aviation English Research Project*.

41. Steven Cushing, "Pilot–Air Traffic Control Communications: It's Not (Only) What You Say, It's How You Say It," *Flight Safety Digest* 14, no. 7 (1995): 1–10.

42. Brett R. C. Molesworth and Dominique Estival, "Miscommunication in General Aviation: The Influence of External Factors on Communication Errors," *Safety Science* 73 (2015): 73–79.

43. Jon Ziomek, *Collision on Tenerife: The How and Why of the World's Worst Aviation Disaster* (New York: Post Hill Press, 2018).

44. Simon Cookson, "Zagreb and Tenerife: Airline Accidents Involving Linguistic Factors," *Australian Review of Applied Linguistics* 32, no. 3 (2009): 1–14.

45. Douglas Kalajian, "Miscommunication That Led to Aviation Disaster," *Chicago Tribune*, April 21, 2002.

46. Malcolm Gladwell, "The Ethnic Theory of Plane Crashes," in *Outliers* (Boston: Little, Brown, 2008), 177–223.

47. Atsushi Tajima, "Fatal Miscommunication: English in Aviation Safety," *World Englishes* 23, no. 3 (2004): 451–70.

BIBLIOGRAPHY

Aamodt, Caitlin. "Collective False Memories: What's behind the 'Mandela Effect'?" *Discover*, February 16, 2017. https://www.discovermagazine.com/mind/collective-false-memories-whats-behind-the-mandela-effect

Abelson, Robert P. "Psychological Status of the Script Concept." *American Psychologist* 36, no. 7 (1981): 715–29.

Acklin, Dina, and Megan H. Papesh. "Modern Speed-Reading Apps Do Not Foster Reading Comprehension." *American Journal of Psychology* 130, no. 2 (2017): 183–99.

Aguilar, Krissy. "SB19 Accused of 'Racism' after Tweeting 'Hello, Negros!' for Concert Tour." *Inquirer.net*, December 27, 2019. https://entertainment.inquirer.net/356469/sb19-accused-of-racism-after-tweeting-hello-negros-for-concert-tour

Akçayır, Murat, Hakan Dündar, and Gökçe Akçayır. "What Makes You a Digital Native? Is It Enough to Be Born after 1980?" *Computers in Human Behavior* 60 (2016): 435–40.

Akiyama, Mitchell. "Silent Alarm: The Mosquito Youth Deterrent and the Politics of Frequency." *Canadian Journal of Communication* 35 (2010): 455–71.

Alacón, Daniel. "How Do You Define a Gang Member?" *New York Times*, May 27, 2015. https://www.nytimes.com/2015/05/31/magazine/how-do-you-define-a-gang-member.html

Albeck-Ripka, Livia. "Missing Apostrophe in Facebook Post Lands a Man in Defamation Court." *New York Times*, October 11, 2021. https://www.nytimes.com/2021/10/11/world/australia/facebook-post-missing-apostrophe-defamation.html

Albert, Saul, and J. P. De Ruiter. "Repair: The Interface between Interaction and Cognition." *Topics in Cognitive Science* 10, no. 2 (2018): 279–313.

Alderson, J. Charles. "Air Safety, Language Assessment Policy, and Policy Implementation: The Case of Aviation English." *Annual Review of Applied Linguistics* 29 (2009): 168–87.

Alderson-Day, Ben, Cesar F. Lima, Samuel Evans, Saloni Krishnan, Pradheep Shanmugalingam, Charles Fernyhough, and Sophie K. Scott. "Distinct Processing of Ambiguous Speech in People with Non-Clinical Auditory Verbal Hallucinations." *Brain* 140, no. 9 (2017): 2475–2489.

Ali, N., A. A. Khan, M. Akunjee, and F. Ahfat. "Using Common Ophthalmologic Jargon in Correspondence Can Lead to Miscommunication." *British Journal of General Practice* (December 2006): 968–69.

Allan, Keith. "The Pragmatics of Connotation." *Journal of Pragmatics* 39, no. 6 (2007): 1047–57.

Allen, Robert. *How to Write Better English*. London: Penguin, 2005.

Allport, Gordon, and Leo Postman. *The Psychology of Rumor*. New York: Henry Holt, 1947.

Alshammari, Khalid Khulaif. "Directness and Indirectness of Speech Acts in Requests among American Native English Speakers and Saudi Native Speakers of Arabic." *English Literature and Language Review* 1, no. 8 (2015): 63–69.

Anderson, Jill. "Misreading Like a Lawyer: Cognitive Bias in Statutory Interpretation." *Harvard Law Review*, 127, no. 6 (2014): 1521–92.

Andrews, Wilson, and Josh Katz. "Language Quiz: Are You on Fleek?" *New York Times*, February 22, 2015.

Apoux, Frédéric, and Sid P. Bacon. "Relative Importance of Temporal Information in Various Frequency Regions for Consonant Identification in Quiet and in Noise." *Journal of the Acoustical Society of America* 116, no. 3 (2004): 1671–80.

Asseraf, Arthur. *Electronic News in Colonial Algeria*. Oxford: Oxford University Press, 2019.

Associated Press. "Deaf Man Attacked because of Sign Language Mixup." March 4, 1995.

Astore, William J. "All the Euphemisms We Use for 'War.'" *The Nation*, April 15, 2016.

Aviezer, Ora. "Bedtime Talk of Three-Year-Olds: Collaborative Repair of Miscommunication." *First Language* 23, no. 1 (2003): 117–39.

Azoulay, Roy, Uri Berger, Hadar Keshet, Paula M. Niedenthal, and Eva Gilboa-Schechtman. "Social Anxiety and the Interpretation of Morphed Facial Expressions Following Exclusion and Inclusion." *Journal of Behavior Therapy and Experimental Psychiatry* 66 (2020): 101–11.

Bacso, Sarah A., and Elizabeth S. Nilsen. "What's That You're Saying? Children with Better Executive Functioning Produce and Repair Communication More Effectively." *Journal of Cognition and Development* 18, no. 4 (2017): 441–64.

Bai, Qiyu, Qi Dan, Zhe Mu, and Maokun Yang. "A Systematic Review of Emoji: Current Research and Future Perspectives." *Frontiers in Psychology* 10 (2019): 2221.

Bakos, Yannis, Florencia Marotta-Wurgler, and David R. Trossen. "Does Anyone Read the Fine Print? Consumer Attention to Standard-Form Contracts." *Journal of Legal Studies* 43, no. 1 (2014): 1–35.

Baron-Cohen, Simon, Daniel C. Bowen, Rosemary J. Holt, Carrie Allison, Bonnie Auyeung, Michael V. Lombardo, Paula Smith, and Meng-Chuan Lai. "The 'Reading the Mind in the Eyes' Test: Complete Absence of Typical Sex Difference in ~400 Men and Women with Autism." *PLoS One* 10, no. 8 (2015): e0136521.

Baron-Cohen, Simon, Therese Jolliffe, Catherine Mortimore, and Mary Robertson. "Another Advanced Test of Theory of Mind: Evidence from Very High Functioning Adults with Autism or Asperger Syndrome." *Journal of Child Psychology and Psychiatry* 38, no. 7 (1997): 813–22.

Barr, Jason, Fred H. Smith, and Sayali J. Kulkarni. "What's Manhattan Worth? A Land Values Index from 1950 to 2014." *Regional Science and Urban Economics* 70 (2018): 1–19.

Barthes, Roland. *Image-Music-Text.* Translated by Stephen Heath. New York: Hill and Wang, 1977.

Bartlett, Frederic C. *Remembering: A Study in Experimental and Social Psychology.* Cambridge: Cambridge University Press, 1932/1997.

Bartlett's Book of Love Quotations. New York: Little, Brown, 1994.

Battistella, Edwin L. "The Syntax of the Double Modal Construction." *Linguistica Atlantica* 17 (1995): 19–44.

Becker, Bernie, and Aaron Lorenzo. "Tax Writers See Peril in Trump's Obamacare Persistence." *Politico*, August 7, 2017. https://www.politico.com/story/2017/08/07/trump-obamacare-congress-tax-reform-241340

Beckwith, Ryan Teague. "Read the Interview Where Gary Johnson Asked What Aleppo Is." *Time*, September 8, 2016.

Behere, Rishikesh V. "Facial Emotion Recognition Deficits: The New Face of Schizophrenia." *Indian Journal of Psychiatry* 57, no. 3 (2015): 229.

Bell, Allan. "Media (Mis)communication on the Science of Climate Change." *Public Understanding of Science 3*, no. 3 (2016): 259–73.

Bennett, Jessica. "I'm Not Mad. That's Just My RBF." *New York Times*, August 1, 2015.

———. "When Your Punctuation Says It All (!)." *New York Times*, March 1, 2015, ST2.

Benson, Robert W. "The End of Legalese: The Game Is Over." *NYU Review of Law and Social Change* 13 (1984): 519–73.

Ben-Ze'ev, Aaron. "The Vindication of Gossip." In *Good Gossip*, edited by Robert F. Goodman and Aaron Ben-Ze'ev, 11–24. Lawrence: University Press of Kansas, 1994.

Berkowitz, Dana. *Botox Nation: Changing the Face of America*. New York: New York University Press, 2017.

Berman, Adrienne. "Reducing Medication Errors through Naming, Labeling, and Packaging." *Journal of Medical Systems* 28, no. 1 (2004): 9–29.

Best, Joel, and Mary M. Hutchinson. "The Gang Initiation Rite as a Motif in Contemporary Crime Discourse." *Justice Quarterly* 13, no. 3 (1996): 383–404.

Bever, Thomas G. "The Cognitive Basis for Linguistic Structures." In *Cognition and the Development of Language*, edited by J. R. Hayes, 1–61. New York: Wiley, 1970.

Biddle, Sam. "Justine Sacco Is Good at Her Job, and How I Came to Peace with Her." *Gawker*, December 20, 2014. https://gawker.com/justine-sacco-is-good-at-her-job-and-how-i-came-to-pea-1653022326

Bisceglio, Paul. "Why Don't Cars Have a 'Sorry' Signal?" *Pacific Standard*, February 3, 2014. https://psmag.com/environment/dont-cars-sorry-signal-73890

Bloom, Harold. *A Map of Misreading*. Oxford: Oxford University Press, 1975.

Bocchino, Anthony J., and David A. Sonenshein. *A Practical Guide to Federal Evidence: Objections, Responses, Rules, and Practice Commentary*. 8th ed. Louisville, CO: National Institute for Trial Advocacy, 2006.

Borrelli, Christopher. "Reagan Used Her, the Country Hated Her. Decades Later, the Welfare Queen of Chicago Refuses to Go Away." *Chicago Tribune*, June 10, 2019.

Boyle, David. *V for Victory: The Wireless Campaign That Defeated the Nazis*. London: The Real Press, 2016.

Bravo, Tony. "How to 'Smize' Like You Mean It." *San Francisco Chronicle*, May 2, 2020.

Brennan, Susan E., Marie K. Huffman, and S. Hannigan. "More Adventures with Dialects: Convergence and Ambiguity Resolution by Partners in Conversation." n.d. https://www.slideplayer.com/slide/4018652

Brewer, William F. "Bartlett's Concept of the Schema and Its Impact on Theories of Knowledge Representation in Contemporary Cognitive Psychology." In *Bartlett, Culture and Cognition*, edited by Akiko Saito, 69–89. Hove: Psychology Press, 2000.

Brighton, Terry. *Hell Riders: The True Story of the Charge of the Light Brigade.* New York: Henry Holt, 2004.

Bromwich, Jonah. "'I Guess I'm Having an Aleppo Moment': Gary Johnson Can't Name a Single Foreign Leader." *New York Times*, September 28, 2016.

Broome, Fiona. *The Mandela Effect—Major Memories, Book 1.* N.p.: New Forest Books, 2019.

Brown, Scott. "Daily Poll: Do You Hear Laurel or Yanny?" *Vancouver Sun*, May 16, 2018.

Bryson, Bill. *The Mother Tongue: English & How It Got That Way.* New York: W. Morrow, 1990.

Bucaria, Chiara. "Lexical and Syntactic Ambiguity as a Source of Humor: The Case of Newspaper Headlines." *Humor* 17, no. 3 (2004): 279–310.

Burnett, Stephanie. "Have You Ever Wondered Why East Asians Spontaneously Make V-Signs in Photos?" *Time*, August 3, 2014. https://time.com/2980357/asia-photos-peace-sign-v-janet-lynn-konica-jun-inoue

Burridge, Kate, and Tonya N. Stebbins. *For the Love of Language: An Introduction to Linguistics.* Cambridge: Cambridge University Press, 2016.

Buss, David M. "The Evolution of Human Intrasexual Competition: Tactics of Mate Attraction." *Journal of Personality and Social Psychology* 54, no. 4 (1988): 616–28.

Byron, Kristin. "Carrying Too Heavy a Load? The Communication and Miscommunication of Emotion by Email." *Academy of Management Review* 33, no. 2 (2008): 309–27.

Calder, Andrew J., Jill Keane, Tom Manly, Reiner Sprengelmeyer, Sophie Scott, Ian Nimmo-Smith, and Andrew W. Young. "Facial Expression Recognition across the Adult Life Span." *Neuropsychologia* 41, no. 2 (2003): 195–202.

Cantor, Allison D., and Elizabeth J. Marsh. "Expertise Effects in the Moses Illusion: Detecting Contradictions with Stored Knowledge." *Memory* 25, no. 2 (2017): 220–30.

Caplan, Jeremy. "Cause of Death: Sloppy Doctors." *Time*, May 15, 2007. http://content.time.com/time/health/article/0,8599,1578074,00.html

Carbon, Claus-Christian. "Wearing Face Masks Strongly Confuses Counterparts in Reading Emotions." *Frontiers in Psychology*, September 25, 2020. https://www.frontiersin.org/articles/10.3389/fpsyg.2020.566886/full

Carey, Mark S. "Nonverbal Openings to Conversation." Paper presented at the meeting of the Eastern Psychological Association, Philadelphia, 1974.

Carlier, J. C., and C. T. Watts. "Roland Barthes's Resurrection of the Author and Redemption of Biography." *The Cambridge Quarterly* 29, no. 4 (2000): 386–93.

Carlisle, Rodney P. *Persian Gulf War*. New York: Facts on File, 2009.

Carrol, Gareth, Jeannette Littlemore, and Margaret Gillon Dowens. "Of False Friends and Familiar Foes: Comparing Native and Non-Native Understanding of Figurative Phrases." *Lingua* 204 (2018): 21–44.

Caucci, Gina M., and Roger J. Kreuz. "Social and Paralinguistic Cues to Sarcasm." *Humor* 25, no. 1 (2012): 1–22.

Caulfield, Philip. "Group of Deaf, Mute Friends Stabbed at Bar after Thug Mistakes Sign Language for Gang Signs." *Daily Mail*, May 1, 2011.

Chengelis, Angelique. "Big Picture: Play Cards Might Look Silly but Are 'Very Important' in Michigan Offense." *Detroit News*, September 19, 2019.

Chipkin, Dana. *Successful Freelance Court Reporting*. Albany, NY: Delmar West Legal Studies, 2001.

Chokshi, Niraj. "That Wasn't Mark Twain: How a Misquotation Is Born." *New York Times*, April 26, 2017.

Clark, Barbara. *Aviation English Research Project: Data Analysis Findings and Best Practice Recommendations*. Gatwick: Civil Aviation Authority, 2017.

Clark, Eve V. *First Language Acquisition*. 2nd ed. Cambridge: Cambridge University Press, 2009.

Clark, Herbert H. *Using Language*. Cambridge: Cambridge University Press, 1996.

Clark, Herbert H., and Susan E. Brennan. "Grounding in Communication." In *Perspectives on Socially Shared Cognition*, edited by Lauren B. Resnick, John M. Levine, and Stephanie D. Teasley, 127–49. Washington, DC: American Psychological Association, 1991.

Clark, Herbert H., and Jean E. Fox Tree. "Using 'Uh' and 'Um' in Spontaneous Speaking." *Cognition* 84, no. 1 (2002): 73–111.

Clark, Herbert H., and Edward F. Schaefer. "Concealing One's Meaning from Overhearers." *Journal of Memory and Language* 26, no. 2 (1987): 209–25.

Cohen, Noam. "Spinning a Web of Lies at Digital Speed." *New York Times*, October 12, 2008, B3.

Cookson, Simon. "Zagreb and Tenerife: Airline Accidents Involving Linguistic Factors." *Australian Review of Applied Linguistics* 32, no. 3 (2009): 22.1–22.14.

Cott, Jonathan. "John Lennon: The Rolling Stone Interview." *Rolling Stone,* November 23, 1968.

Cotton, Gayle. *Say Anything to Anyone, Anywhere: 5 Keys to Successful Cross-Cultural Communication.* Hoboken, NJ: Wiley, 2013.

Crowe, S. F., J. Barot, S. Caldow, J. d'Aspromonte, J. Dell'Orso, A. Di Clemente, K. Hanson, M. Kellett, S. Makhlota, B. McIvor, L. McKenzie, R. Norman, A. Thiru, M. Twyerould, and S. Sapega. "The Effect of Caffeine and Stress on Auditory Hallucinations in a Non-Clinical Sample." *Personality and Individual Differences* 50, no. 5 (2011): 626–30.

Cumming-Bruce, Nick. "Myanmar Generals Should Face Genocide Charges over Rohingya, U.N. Says." *New York Times,* August 27, 2018.

———. "U.S. Steps Up Criticism of China for Detentions in Xinjiang." *New York Times,* March 13, 2019.

Cunningham, Michael R. "Reactions to Heterosexual Opening Gambits: Female Selectivity and Male Responsiveness." *Personality and Social Psychology Bulletin* 15, no. 1 (1989): 27–41.

Curry, Jack. "Mets Follow the Wrong Sign to Another Defeat." *New York Times,* June 14, 1991, B11, B13.

Cushing, Steven. "Pilot–Air Traffic Control Communications: It's Not (Only) What You Say, It's How You Say It." *Flight Safety Digest* 14, no. 7 (1995): 1–10.

Daily Mail. "Jail for Knifeman Who Stabbed Friend to Death in Row over Text Message." April 1, 2011.

Daneman, Meredyth, and Murray Stainton. "The Generation Effect in Reading and Proofreading: Is It Easier or Harder to Detect Errors in One's Own Writing?" *Reading and Writing* 5, no. 3 (1993): 297–313.

Davich, Jerry. "Text Messages Can Easily Get Lost in Translation." *Chicago Tribune,* June 29, 2017.

Davies, Mark. "Expanding Horizons in Historical Linguistics with the 400-Million Word Corpus of Historical American English." *Corpora* 7, no. 2 (2012): 121–57.

Davis, William. "Really, This Jargon is Getting Out of Hand." *New York Times,* March 25, 1973, 3.

Dawes, Piers, Richard Emsley, Karen J. Cruickshanks, David R. Moore, Heather Fortnum, Mark Edmondson-Jones, Abby McCormack, and Kevin

J. Munro. "Hearing Loss and Cognition: The Role of Hearing Aids, Social Isolation and Depression." *PLoS One* 10, no. 3 (2015): e0119616.

Dazzi, Carla, and Luigi Pedrabissi. "Graphology and Personality: An Empirical Study on Validity of Handwriting Analysis." *Psychological Reports* 105, no. 3, suppl. (2009): 1255–68.

DeJohn, Irving, and Helen Kennedy. "'Jeremy Lin Headline Slur Was 'Honest Mistake,' Fired ESPN Editor Anthony Federico Claims." *Daily News*, February 20, 2012. https://www.nydailynews.com/sports/basketball/knicks/jeremy-lin-slur-honest-mistake-fired-espn-editor-anthony-federico-claims-article-1.1025566

De Jong, Peter F., Daniëlle J. L. Bitter, Margot Van Setten, and Eva Marinus. "Does Phonological Recoding Occur during Silent Reading, and Is It Necessary for Orthographic Learning?" *Journal of Experimental Child Psychology* 104, no. 3 (2009): 267–82.

Dewaele, Jean-Marc. "Interpreting Grice's Maxim of Quantity: Interindividual and Situational Variation in Discourse Styles of Non Native Speakers." In *Cognition in Language Use: Selected Papers from the 7th International Pragmatics Conference*, vol. 1, edited by E. Németh, 85–99. Antwerp: International Pragmatics Association, 2001.

Dickey, Beth. "Spacecraft Is Launched to Look for Water on Mars." *New York Times*, December 12, 1998, A10.

DiFonzo, Nicholas, and Prashant Bordia. "Rumor, Gossip, and Urban Legends." *Diogenes* 213 (2007): 19–35.

Dingemanse, Mark, Seán G. Roberts, Julija Baranova, Joe Blythe, Paul Drew, Simeon Floyd, Rosa S. Gisladottir, Kobin H. Kendrick, Stephen C. Levinson, Elizabeth Manrique, Giovanni Rossi, and N. J. Enfield. "Universal Principles in the Repair of Communication Problems." *PLoS One* 10, no. 9 (2015): e0136100.

Dodell-Feder, David, Kerry J. Ressler, and Laura T. Germine. "Social Cognition or Social Class and Culture? On the Interpretation of Differences in Social Cognitive Performance." *Psychological Medicine* 50, no. 1 (2020): 133–45.

Dorsen, David. "Is Gorsuch an Originalist? Not So Fast." *Washington Post*, March 17, 2017.

Dressler, Joshua. "Rethinking Heat of Passion: A Defense in Search of a Rationale." *Journal of Criminal Law & Criminology* 73 (1982): 421–70.

Drewnowski, Adam, and Alice F. Healy. "Detection Errors on *the* and *and*: Evidence for Reading Units Larger Than the Word." *Memory & Cognition* 5, no. 6 (1977): 636–47.

Dreyer, Benjamin. *Dreyer's English: An Utterly Correct Guide to Clarity and Style*. New York: Random House, 2019.

Driver, Russell W., M. Ronald Buckley, and Dwight D. Frink. "Should We Write Off Graphology?" *International Journal of Selection and Assessment* 4, no. 2 (1996): 78–86.

Ducharme, Jamie. "No, Teenagers Are Not Growing 'Skull Horns' because of Smartphones." *Time*, June 21, 2019. https://time.com/5611036/teenagers -skull-horns

Duprey, Rich. "Drinkers Are Confusing Corona Beer with the Coronavirus." *The Motley Fool*, February 5, 2020. https://www.fool.com/investing/2020/02/05/drinkers-are-confusing-corona-beer-with-the-corona.aspx

Ego, Michael M. "'Chink in the Armor': Is It a Racist Cliché?" *AP News*, October 28, 2018. https://apnews.com/4722f93c10a141d39ddb0a65aef16c49

Ekman, Paul, Richard J. Davidson, and Wallace V. Friesen. "The Duchenne Smile: Emotional Expression and Brain Physiology: II." *Journal of Personality and Social Psychology* 58, no. 2 (1990): 342–53.

Elliott, Chris. "The Reader's Editor on . . . How Headlines Can Be More Easily Misunderstood Online." *The Guardian*, September 16, 2012.

Ellis, Edward Robb. *The Epic of New York City: A Narrative History*. New York: Basic Books, 1966.

Engelhardt, Paul E., Karl G. D. Bailey, and Fernanda Ferreira. "Do Speakers and Listeners Observe the Gricean Maxim of Quantity?" *Journal of Memory and Language* 54, no. 4 (2006): 554–73.

Erickson, Thomas D., and Mark E. Mattson. "From Words to Meaning: A Semantic Illusion." *Journal of Verbal Learning and Verbal Behavior* 20, no. 5 (1981): 540–51.

Eschner, Kat. "Where the Myth of JFK's 'Jelly Donut' Mistake Came From." *Smithsonian*, June 26, 2017.

Evans, Vyvyan. *The Emoji Code: The Linguistics behind Smiley Faces and Scaredy Cats*. New York: Picador, 2017.

Facebook Community Standards, Part III, Section 15. "Sexual Solicitation." August 2019. https://www.facebook.com/communitystandards/sexual_solicitation

Fadaee, Elaheh. "Symbols, Metaphors and Similes in Literature: A Case Study of 'Animal Farm.'" *Journal of English and Literature* 2, no. 2 (2011): 19–27.

Fagan, Mary K. "Toddlers' Persistence When Communication Fails: Response Motivation and Goal Substitution." *First Language* 28, no. 1 (2008): 55–69.

Farmer, Sam. "Ask Farmer: How Many Plays in an NFL Playbook, and How Many Do Teams Run in a Game?" *Los Angeles Times*, January 3, 2016.

Feldman Barrett, Lisa, Ralph Adolphs, Stacy Marsella, Aleix M. Martinez, and Seth D. Pollak. "Emotional Expressions Reconsidered: Challenges to Inferring Emotion from Human Facial Movements." *Psychological Science in the Public Interest* 20, no. 1 (2019): 1–68.

Finneran, Kevin. "Can the Public Be Trusted?" *Issues in Science and Technology* 34, no. 4 (2018): 17–18.

Fisher, Marc. "Donald Trump Doesn't Read Much. Being President Probably Wouldn't Change That." *Washington Post*, July 17, 2016.

Fleming, John H., John M. Darley, James L. Hilton, and Brian A. Kojetin. "Multiple Audience Problem: A Strategic Communication Perspective on Social Perception." *Journal of Personality and Social Psychology* 58, no. 4 (1990): 593–609.

Floyd, Simeon, Elizabeth Manrique, Giovanni Rossi, and Francisco Torreira. "Timing of Visual Bodily Behavior in Repair Sequences: Evidence from Three Languages." *Discourse Processes* 53, no. 3 (2016): 175–204.

Ford, Judith M., Thomas Dierks, Derek J. Fisher, Christoph S. Herrmann, Daniela Hubl, Jochen Kindler, Thomas Koenig, Daniel H. Mathalon, Kevin M. Spencer, Werner Strik, and Remko van Lutterveld. "Neurophysiological Studies of Auditory Verbal Hallucinations." *Schizophrenia Bulletin* 38, no. 4 (2012): 715–23.

Fortin, Jacey. "Kevin Hart Steps Down as Oscars Host after Criticism over Homophobic Tweets." *New York Times*, December 6, 2018.

Foster, Eric K. "Research on Gossip: Taxonomy, Methods, and Future Directions." *Review of General Psychology* 8, no. 2 (2004): 78–99.

Fowler, H. W. *A Dictionary of Modern English Usage*. Oxford: Oxford University Press, 1965.

Fox Tree, Jean E., J. Trevor D'Arcey, Alicia A. Hammond, and Alina S. Larson. "The Sarchasm: Sarcasm Production and Identification in Spontaneous Conversation." *Discourse Processes* 57, no. 5–6 (2020): 1–27.

Francis, Peter, Jr. "The Beads That Did 'Not' Buy Manhattan Island." *New York History* 78, no. 4 (October 1997): 411–28.

Freedman, Adam. "Clause and Effect." *New York Times*, December 16, 2007.

French, Aaron. "The Mandela Effect and New Memory." *Correspondences: Journal for the Study of Esotericism* 6, no. 2 (2018): 201–33.

Freud, Sigmund. *The Psychopathology of Everyday Life*. New York: Norton, 1901/1989.

Friedman, Thomas L. "Too Good to Check." *New York Times*, November 17, 2010, A33.

Frisby, Brandi N., Megan R. Dillow, Shelbie Gaughan, and John Nordlund. "Flirtatious Communication: An Experimental Examination of Perceptions of Social-Sexual Communication Motivated by Evolutionary Forces." *Sex Roles* 64, no. 9–10 (2011): 682–94.

Frishberg, Hannah. "'Sexual' Use of Eggplant and Peach Emojis Banned on Facebook, Instagram." *New York Post*, October 20, 2019.

Fumento, Michael. "'Road Rage' versus Reality." *Atlantic Monthly* 282, no. 2 (August 1998): 12–17.

Gambrell, Linda B. "The Occurrence of Think-Time during Reading Comprehension Instruction." *Journal of Educational Research* 77, no. 2 (1983): 77–80.

Gardiner, Richard. "The Civil War Origin of Coca-Cola in Columbus, Georgia." *Muscogee Genealogical Society* 2012: 21–24.

Gardner, Howard. "Getting Acquainted with Jean Piaget." *New York Times*, January 3, 1979, C1.

Garner, Bryan. *Garner's Modern English Usage*. 4th ed. Oxford: Oxford University Press, 2016.

Gary, Jerry. "No. 2 House Leader Refers to Colleague with Anti-Gay Slur." *New York Times*, January 28, 1995, 1.

Gernsbacher, Morton Ann, and David J. Hargreaves. "Accessing Sentence Participants: The Advantage of First Mention." *Journal of Memory and Language* 27, no. 6 (1988): 699–717.

Gernsbacher, Morton Ann, Adam R. Raimond, M. Theresa Balinghasay, and Jilana S. Boston. "'Special Needs' Is an Ineffective Euphemism." *Cognitive Research: Principles and Implications* 1, no. 1 (2016): 1–13.

Ghosh, Debanjan, Alexander R. Fabbri, and Smaranda Muresan. "Sarcasm Analysis Using Conversation Context." *Computational Linguistics* 44, no. 4 (2018): 755–92.

Gibbs, Raymond W., Jr. "Irony in Talk among Friends." *Metaphor and Symbol* 15, no. 1–2 (2000): 5–27.

Gibbs, Raymond W., Jr., and Jennifer E. O'Brien. "Idioms and Mental Imagery: The Metaphorical Motivation for Idiomatic Meaning." *Cognition* 36, no. 1 (1990): 35–68.

Gil-Lopez, Teresa, Cuihua Shen, Grace A. Benefield, Nicholas A. Palomares, Michal Kosinski, and David Stillwell. "One Size Fits All: Context Collapse, Self-Presentation Strategies and Language Styles on Facebook." *Journal of Computer-Mediated Communication* 23, no. 3 (2018): 127–45.

Gilovich, Thomas, Victoria Husted Medvec, and Kenneth Savitsky. "The Spotlight Effect in Social Judgment: An Egocentric Bias in Estimates of the

Salience of One's Own Actions and Appearance." *Journal of Personality and Social Psychology* 78, no. 2 (2000): 211–22.

Gilovich, Thomas, Kenneth Savitsky, and Victoria Husted Medvec. "The Illusion of Transparency: Biased Assessments of Others' Ability to Read One's Emotional States." *Journal of Personality and Social Psychology* 75, no. 2 (1998): 332–46.

Gladwell, Malcolm. "The Ethnic Theory of Plane Crashes." In *Outliers*, 177–223. Boston: Little, Brown, 2008.

Goldberg, Daniel M. "What Comprises a 'Lascivious Exhibition of the Genitals or Pubic Area'? The Answer, My Friend, Is *Blouin* in the Wind." *Military Law Review*, 224 (2016): 425–80.

Goldman, A. "What's Up with Chuck Grassley's Twitter Feed?" *Digg*, April 16, 2015. https://digg.com/2015/chuck-grassley-twitter

Goldman, Eric. "Emojis and the Law." *Washington Law Review* 93 (2018): 1227–91.

Goldstein, Tom. "Lawyers Now Confuse Even the Same Aforementioned." *New York Times*, April 1, 1977, 23.

Goluboff, Risa. "The Forgotten Law That Gave Police Nearly Unlimited Power." *Time*, February 1, 2016.

Gorman, Glen, and Christian H. Jordan. "'I Know You're Kidding:' Relationship Closeness Enhances Positive Perceptions of Teasing." *Personal Relationships* 22, no. 2 (2015): 173–87.

Gorvett, Zaria. "How Modern Life Is Transforming the Human Skeleton." *BBC Future*, June 13, 2019.

Grady, Denise. "About the Idea That You're Growing Horns from Looking Down at Your Phone . . .". *New York Times*, June 20, 2019.

Grayson, Mara L. *Teaching Racial Literacy: Reflective Practices for Critical Writing*. Lanham, MD: Rowman & Littlefield, 2018.

Green, Emma. "The Origins of Office Speak: What Corporate Buzzwords Reveal about the History of Work." *The Atlantic*, April 24, 2014.

Grice, H. P. "Logic and Conversation." In *Syntax and Semantics 3: Speech Acts*, edited by P. Cole and J. L. Morgan, 41–58. New York: Academic Press, 1975.

Gross, Daniel. "That's What She Said: The Rise and Fall of the 2000s' Best Bad Joke." *The Atlantic*, January 24, 2014.

Grossman, Samantha. "Deaf Man Stabbed after Sign Language Mistaken for Gang Signs." *Time*, January 15, 2013.

Guéguen, Nicolas, Jacques Fischer-Lokou, Liv Lefebvre, and Lubomir Lamy. "Women's Eye Contact and Men's Later Interest: Two Field Experiments." *Perceptual and Motor Skills* 106, no. 1 (2008): 63–66.

Gunraj, Danielle N., April M. Drumm-Hewitt, Erica M. Dashow, Sri Siddhi N. Upadhyay, and Celia M. Klin. "Texting Insincerely: The Role of the Period in Text Messaging." *Computers in Human Behavior* 55 (2016): 1067–75.

Haber, Matt. "Should Grown Men Use Emoji?" *New York Times*, April 3, 2015, D27.

Halley, Janet. "The Move to Affirmative Consent." *Signs: Journal of Women in Culture and Society* 42, no. 1 (2016): 257–79.

Halperin, Alex. "Did Urban Outfitters Mean to Put a Gang Sign on a Shirt?" *Salon*, July 18, 2013.

Halvorson, Heidi Grant. *No One Understands You and What to Do about It*. Boston: Harvard Business Review Press, 2015.

Hamilton, Cheryl. *Communicating for Results: A Guide for Business and the Professions*. 10th ed. Boston: Wadsworth Cengage, 2014.

Hancock, Jason. "Hawley's Critique of 'Cosmopolitan Elite' Earns Rebuke from Missouri Jewish Leaders." *Kansas City Star*, July 19, 2019. https://www.kansascity.com/news/politics-government/article232902747.html

Harland, David M. *Apollo 12—On the Ocean of Storms*. New York: Springer, 2011.

Harms, Madeline B., Alex Martin, and Gregory L. Wallace. "Facial Emotion Recognition in Autism Spectrum Disorders: A Review of Behavioral and Neuroimaging Studies." *Neuropsychology Review* 20, no. 3 (2010): 290–322.

Harrison, Hy. "The Origin of 'Yankee.'" *The Academy and Literature* 1913: 222–23.

Hartshorne, Joshua K., and Laura T. Germine. "When Does Cognitive Functioning Peak? The Asynchronous Rise and Fall of Different Cognitive Abilities across the Life Span." *Psychological Science* 26, no. 4 (2015): 433–43.

Haskell, Robert E. "Unconscious Linguistic Referents to Race: Analysis and Methodological Frameworks." *Discourse & Society* 20, no. 1 (2009): 59–84.

Haugh, Michael. "'Just Kidding': Teasing and Claims to Non-Serious Intent." *Journal of Pragmatics* 95 (2016): 120–36.

Healey, Patrick G. T., Gregory J. Mills, Arash Eshghi, and Christine Howes. "Running Repairs: Coordinating Meaning in Dialogue." *Topics in Cognitive Science* 10, no. 2 (2018): 367–88.

Hébert, Nathaniel. "Sex IN THE City; Carrie Bradshaw and the Mandela Effect." *Medium*, May 23, 2018. https://medium.com/@nathanielhebert/sex-in-the-city-carrie-bradshaw-and-the-mandela-effect-bead3a66b7bf

Heim, Joe. "Once All but Left for Dead, Is Cursive Handwriting Making a Comeback?" *Washington Post*, July 26, 2016.

Heitz, Richard P. "The Speed-Accuracy Tradeoff: History, Physiology, Methodology, and Behavior." *Frontiers in Neuroscience* 8 (2014): 150.

Hemforth, Barbara, Lars Konieczny, Christoph Scheepers, Saveria Colonna, Sarah Schimke, Peter Baumann, and Joël Pynte. "Language Specific Preferences in Anaphor Resolution: Exposure or Gricean Maxims?" In *32nd Annual Conference of the Cognitive Science Society* 2010: 2218-23. hal -01015048

Hemmens, Craig, Kathryn E. Scarborough, and Rolando V. Del Carmen. "Grave Doubts about 'Reasonable Doubt': Confusion in State and Federal Courts." *Journal of Criminal Justice* 25, no. 3 (1997): 231-54.

Henderson-Sellers, Ann. "Climate Whispers: Media Communication about Climate Change." *Climatic Change* 40, no. 3-4 (1998): 421-56.

Hendrickson, John. "Gary Johnson Is Ready to Talk about Aleppo." *Esquire*, February 23, 2018.

Henningsen, David Dryden. "Flirting with Meaning: An Examination of Miscommunication in Flirting Interactions." *Sex Roles* 50, no. 7-8 (2004): 481-89.

Herbst, Abigail. "How the Term 'Social Justice Warrior' Became an Insult." *Foundation for Economic Education*, August 13, 2018. https://fee.org/articles/how-the-term-social-justice-warrior-became-an-insult

Hibbert, Christopher. *The Destruction of Lord Raglan: A Tragedy of the Crimean War, 1854-55*. London: Longmans, 1961.

High, Rob, and Tanmay Bakshi. *Cognitive Computing with IBM Watson*. Birmingham: Packt Publishing, 2019.

Holler, Anke, and Katja Suckow, eds. *Empirical Perspectives on Anaphora Resolution*. Berlin: Walter de Gruyter, 2016.

Holsinger, M. Paul, ed. *War and American Popular Culture: A Historical Encyclopedia*. Westport, CT: Greenwood Press, 1999.

Hombach, Stella Marie. "From behind the Coronavirus Mask, an Unseen Smile Can Still Be Heard." *Scientific American*, June 1, 2020. https://www.scientificamerican.com/article/from-behind-the-coronavirus-mask-an-unseen-smile-can-still-be-heard

Horowitz, Irwin A. "Reasonable Doubt Instructions: Commonsense Justice and Standard of Proof." *Psychology, Public Policy, and Law* 3, no. 2-3 (1997): 285-302.

Houghton, Kenneth J., Sri Siddhi N. Upadhyay, and Celia M. Klin. "Punctuation in Text Messages May Convey Abruptness. Period." *Computers in Human Behavior* 80 (2018): 112–21.

House, Juliane, Gabriele Kasper, and Steven Ross. "Misunderstanding Talk." In *Misunderstanding in Social Life: Discourse Approaches to Problematic Talk*, edited by Juliane House, Gabriele Kasper, and Steven Ross, 1–21. Harlow: Pearson Education, 2003.

Houston, Rick, and Milt Heflin. *Go Flight! The Unsung Heroes of Mission Control, 1965–1992*. Lincoln: University of Nebraska Press, 2015.

Howard, John W., III. "'Tower, Am I Cleared to Land?': Problematic Communication in Aviation Discourse." *Human Communication Research* 34, no. 3 (2008): 370–91.

Huber, Jonathan D., and Edward S. Herold. "Sexually Overt Approaches in Singles Bars." *Canadian Journal of Human Sexuality* 15, no. 3/4 (2006): 133–46.

Hurley, Peter. *The Headshot: The Secrets to Creating Amazing Headshot Portraits*. San Francisco: New Riders Press, 2016.

Itzkoff, Dave. "Tony Lives, or Doesn't." *New York Times*, August 29, 2014, C3.

Ivanko, Stacey L., Penny M. Pexman, and Kara M. Olineck. "How Sarcastic Are You? Individual Differences and Verbal Irony." *Journal of Language and Social Psychology* 23, no. 3 (2004): 244–71.

Jonakait, Randolph N., Harold Baer Jr., E. Stewart Jones Jr., and Edward J. Imwinkelried. *New York Evidentiary Foundations*. 2nd ed. Charlottesville, VA: The Michie Company, 1998.

Jones, Taylor, Jessica Rose Kalbfeld, Ryan Hancock, and Robin Clark. "Testifying while Black: An Experimental Study of Court Reporter Accuracy in Transcription of African American English." *Language* 95, no. 2 (2019): e216–52.

Kaczynski, Andrew. "Senator Grassley Explains His Odd Dairy Queen Tweet." *BuzzFeed*, November 7, 2014. https://www.buzzfeednews.com/article/andrewkaczynski/senator-grassley-explains-his-odd-dairy-queen-tweet

Kalajian, Douglas. "Miscommunication That Led to Aviation Disaster." *Chicago Tribune*, April 21, 2002.

Katz, Josh, Jonathan Corum, and Jon Huang. "We Made a Tool so You Can Hear Both Yanny and Laurel." *New York Times*, May 16, 2018.

Katzman, Marilyn. "Baffled by Office Buzzwords." *New York Times*, July 4, 2015.

Kavé, Gitit, Ariel Knafo, and Asaf Gilboa. "The Rise and Fall of Word Retrieval across the Lifespan." *Psychology and Aging* 25, no. 3 (2010): 719–24.

Kazin, Michael. *The Populist Persuasion: An American History.* Ithaca, NY: Cornell University Press, 1998.

Kelly, Henry Ansgar. "Rule of 'Thumb' and the Folklaw of the Husband's Stick." *Journal of Legal Education* 44 (1994): 341–65.

Kelly, Lynne, James A. Keaten, Bonnie Becker, Jodi Cole, Lea Littleford, and Barrett Rothe. "'It's the American Lifestyle!': An Investigation of Text Messaging by College Students." *Qualitative Research Reports in Communication* 13, no. 1 (2012): 1–9.

Kelly, Lynne, and Aimee E. Miller-Ott. "Perceived Miscommunication in Friends' and Romantic Partners' Texted Conversations." *Southern Communication Journal* 83, no. 4 (2018): 267–80.

Kelly, Nataly, and Jost Zetzsche. *Found in Translation: How Language Shapes Our Lives and Transforms the World.* New York: Perigee, 2012.

Kelsey, N. G. N. *Games, Rhymes, and Wordplay of London Children.* London: Palgrave Macmillan, 2019.

Keltner, Dacher, Randall C. Young, Erin A. Heerey, Carmen Oemig, and Natalie D. Monarch. "Teasing in Hierarchical and Intimate Relations." *Journal of Personality and Social Psychology* 75, no. 5 (1998): 1231–47.

Keltner, Dacher, Lisa Capps, Ann M. Kring, Randall C. Young, and Erin A. Heerey. "Just Teasing: A Conceptual Analysis and Empirical Review." *Psychological Bulletin* 127, no. 2 (2001): 229–48.

Kemper, Susan. "Elderspeak: Speech Accommodations to Older Adults." *Aging and Cognition* 1, no. 1 (1994): 17–28.

Kendrick, Leslie. "A Test for Criminally Instructional Speech." *Virginia Law Review* 91 (2005): 1973–2021.

Kennedy, Pagan. "Who Made That Emoticon?" *New York Times*, November 23, 2012.

Kentner, Gerrit. "Rhythmic Segmentation in Auditory Illusions: Evidence from Cross-Linguistic Mondegreens." In *Proceedings of the 18th International Congress of Phonetic Sciences.* Glasgow: University of Glasgow, 2015.

Kerr, Orin S. "The Decline of the Socratic Method at Harvard." *Nebraska Law Review* 78 (1999): 113–34.

Keysar, Boaz. "Communication and Miscommunication: The Role of Egocentric Processes." *Intercultural Pragmatics* 4, no. 1 (2007): 71–84.

Keysar, Boaz, and Anne S. Henly. "Speakers' Overestimation of Their Effectiveness." *Psychological Science* 13, no. 3 (2002): 207–12.

Kilpatrick, James. J. *The Ear Is Human: A Handbook of Homophones and Other Confusions*. Kansas City, MO: Andrews, McMeel & Parker, 1985.

Kimball, John. "Seven Principles of Surface Structure Parsing in Natural Language." *Cognition* 2 (1973): 12–47.

Kirkpatrick, David D. "Speaking in the Tongue of Evangelicals." *New York Times*, October 17, 2004.

Kirsch, Adam. "Should an Author's Intentions Matter?" *New York Times*, March 10, 2015, 31.

Kirschner, Paul A., and Pedro De Bruyckere. "The Myths of the Digital Native and the Multitasker." *Teaching and Teacher Education* 67 (2017): 135–42.

Kitsap Sun. "Road Rage: Man Shot When Trying to Apologize." December 29, 1997. https://products.kitsapsun.com/archive/1997/12-29/0020_road_rage__man_shot_when_trying_t.html

Klee, Miles. "Senator Orrin Hatch and the Origin of 'Shooting One's Wad.'" *Mel Magazine*, August 2017. https://melmagazine.com/en-us/story/senator-orrin-hatch-and-the-origin-of-shooting-ones-wad

Knowlton, Brian, and International Herald Tribune. "No Radio Contact with NASA Spacecraft: Orbiter Believed Lost upon Arrival at Mars." *New York Times*, September 24, 1999.

Koblin, John. "After Racist Tweet, Roseanne Barr's Show Is Canceled by ABC." *New York Times*. May 29, 2018.

Kowalski, Robin M. "'I Was Only Kidding!': Victims' and Perpetrators' Perceptions of Teasing." *Personality and Social Psychology Bulletin* 26, no. 2 (2000): 231–41.

Kreuz, Roger. *Irony and Sarcasm*. Cambridge, MA: MIT Press, 2020.

Kreuz, Roger J., and Richard M. Roberts. "Two Cues for Verbal Irony: Hyperbole and the Ironic Tone of Voice." *Metaphor and Symbol* 10, no. 1 (1995): 21–31.

Kruger, Justin, Nicholas Epley, Jason Parker, and Zhi-Wen Ng. "Egocentrism over E-Mail: Can We Communicate as Well as We Think?" *Journal of Personality and Social Psychology* 89, no. 6 (2005): 925–36.

Kruger, Justin, Cameron L. Gordon, and Jeff Kuban. "Intentions in Teasing: When 'Just Kidding' Just Isn't Good Enough." *Journal of Personality and Social Psychology* 90, no. 3 (2006): 412–25.

Krumhuber, Eva G., and Antony S. R. Manstead. "Can Duchenne Smiles Be Feigned? New Evidence on Felt and False Smiles." *Emotion* 9, no. 6 (2009): 807–20.

Krzeszowski, Tomasz P. "The Bible Translation Imbroglio." In *Cultural Conceptualization in Translation and Language Applications*, edited by Barbara

Lewandwoska-Tomaszczyk, 15–33. Cham: Springer Nature Switzerland, 2020.

Kurzon, Dennis. "Towards a Typology of Silence." *Journal of Pragmatics* 39, no. 10 (2007): 1673–88.

La France, Betty H. "What Verbal and Nonverbal Communication Cues Lead to Sex: An Analysis of the Traditional Sexual Script." *Communication Quarterly* 58, no. 3 (2010): 297–318.

Lakoff, George, and Mark Johnson. *Metaphors We Live By*. Chicago: University of Chicago Press, 1980.

Lane, K. Maria D. *Geographies of Mars: Seeing and Knowing the Red Planet*. Chicago: University of Chicago Press, 2011.

Laubert, Christoph, and Jennifer Parlamis. "Are You Angry (Happy, Sad) or Aren't You? Emotion Detection Difficulty in Email Negotiation." *Group Decision and Negotiation* 28, no. 2 (2019): 377–413.

LeBlanc, Thomas W., Ashley Hesson, Andrew Williams, Chris Feudtner, Margaret Holmes-Rovner, Lillie D. Williamson, and Peter A. Ubel. "Patient Understanding of Medical Jargon: A Survey Study of U.S. Medical Students." *Patient Education and Counseling* 95, no. 2 (2014): 238–42.

Lebra, Takie Sugiyama. "The Cultural Significance of Silence in Japanese Communication." *Multilingua* 6, no. 4 (1987): 343–58.

Lederer, Richard. "Curious Contronyms." *Word Ways*, February 1978, 27–28.

Leithauser, Brad. "Unusable Words." *The New Yorker*, October 14, 2013.

Leonidas, Leonardo L. "Death by Bad Handwriting." *Philippine Daily Inquirer*, October 27, 2014. https://opinion.inquirer.net/79623/death-by-bad-handwriting

Levelt, Willem J. M. "Monitoring and Self-Repair in Speech." *Cognition* 14, no. 1 (1983): 41–104.

Levenson, Michael. "Military Says Hand Gestures at Game Were Not White Power Signs." *New York Times*, December 20, 2019.

Levett, Benjamin A. *Through the Customs Maze: A Popular Exposition and Analysis of the United States Customs Tariff Administrative Laws*. New York: Customs Maze Publishing, 1923.

Li, Chengsheng, Ruilan Lu, and D. Li Davis. "Let 500,000,000 Bottles Be Broken All Over China: Underground Dissidents Are Hoping the Small Gesture Will Breathe New Life into the Pro-Democracy Movement." *Los Angeles Post*, April 18, 1990.

Liberman, Mark. "Egg Corns: Folk Etymology, Malapropism, Mondegreen, ???." *Language Log*, September 23, 2003. http://itre.cis.upenn.edu/~myl/languagelog/archives/000018.html

Liu, Jiangang, Jun Li, Lu Feng, Ling Li, Jie Tian, and Kang Lee. "Seeing Jesus in Toast: Neural and Behavioral Correlates of Face Pareidolia." *Cortex* 53 (2014): 60–77.

Lively, Mathew W. *Calamity at Chancellorsville: The Wounding and Death of Confederate General Stonewall Jackson*. El Dorado Hills, CA: Savas Beatie, 2013.

Loveman, Kate. *Reading Fictions, 1660–1740: Deception in English Literary and Political Culture*. Aldershot: Ashgate, 2008.

Lubertozzi, Alex. "Introduction." *Oliver Twist*. Oak Park, IL: Top Five Books, 2019.

Madison, James. "Concerning the Difficulties of the Convention in Devising a Proper Form of Government." *Federalist Paper No. 37*, 1788.

Madison, Jillian. *Damn You, Autorrect! Awesomely Embarrassing Text Messages You Didn't Mean to Send*. New York: Hyperion, 2011.

Magliozzi, Tom, and Ray Magliozzi. "The Horn and Lights Say It All in Basic Language of the Road." *Orlando Sentinel*, April 19, 2001.

Mahan, Logan. "Youthsplaining: You've Been Texting the Word 'Okay' Wrong." *InsideHook*, July 31, 2019. https://www.insidehook.com/article/advice/the-difference-between-texting-k-ok-kk-explained

Maijer, K., M. J. H. Begemann, S. J. M. C. Palmen, S. Leucht, and I. E. C. Sommer. "Auditory Hallucinations across the Lifespan: A Systematic Review and Meta-Analysis." *Psychological Medicine* 48, no. 6 (2018): 879–88.

Makarechi, Kia. "'Pointergate' Is the Most Pathetic News Story of the Week." *Vanity Fair*, November 7, 2014.

Markoff, John. "Computer Wins on 'Jeopardy!': Trivial, It's Not." *New York Times*, February 16, 2011.

Marshall, Ashley. "Daniel Defoe as Satirist." *Huntington Library Quarterly* 70, no. 4 (2007): 553–76.

Marwick, Alice E., and danah boyd. "I Tweet Honestly, I Tweet Passionately: Twitter Users, Context Collapse, and the Imagined Audience." *New Media & Society* 13, no. 1 (2011): 114–33.

Massaro, Dominic W., and Alexandra Jesse. "Read My Lips: Speech Distortions in Musical Lyrics Can Be Overcome (Slightly) by Facial Information." *Speech Communication* 51, no. 7 (2009): 604–21.

McCarl, Ryan. "Incoherent and Indefensible: An Interdisciplinary Critique of the Supreme Court's Void-for-Vagueness Doctrine." *Hastings Constitutional Law Quarterly* 42 (2014): 73–94.

McConchie, Alan. "Pop vs Soda." http://www.popvssoda.com

McCulloch, Gretchen. *Because Internet: Understanding the New Rules of Language*. New York: Riverhead Books, 2019.

McGrath, John, Sukanta Saha, David Chant, and Joy Welham. "Schizophrenia: A Concise Overview of Incidence, Prevalence, and Mortality." *Epidemiologic Reviews* 30, no. 1 (2008): 67–76.

McNeil, Peter. "'That Doubtful Gender': Macaroni Dress and Male Sexualities." *Fashion Theory* 3, no. 4 (1999): 411–47.

McSweeney, Michelle A. *The Pragmatics of Text Messaging: Making Meaning in Messages*. New York: Routledge, 2018.

Mehr, Samuel A. "Miscommunication of Science: Music Cognition Research in the Popular Press." *Frontiers in Psychology* 6 (2015): 988.

Memmott, Mark. "'Eggcorns': The Gaffes That Spread Like Wildflowers." *NPR Weekend Edition Saturday*, May 30, 2015. https://www.npr.org/sections/thetwo-way/2015/05/30/410504851/eggcorns-the-gaffes-that-spread-like-wildflowers

Merriam-Webster. "Bimonthly." https://www.merriamwebster.com/dictionary/bimonthly

———. "Did We Change the Definition of Literally?" https://www.merriam-webster.com/words-at-play/misuse-of-literally

Metz, Cade. "One Genius' Lonely Crusade to Teach a Computer Common Sense." *Wired*, March 24, 2016.

Mikkelson, David. "Did Barack Obama Say He Had Visited 57 (Islamic) States?" *Snopes*, June 19, 2008. https://www.snopes.com/fact-check/57-states

Miller, George A., Eugene Galanter, and Karl H. Pribram. *Plans and the Structure of Behavior*. New York: Henry Holt, 1960.

Miller, Hannah, Daniel Kluver, Jacob Thebault-Spieker, Loren Terveen, and Brent Hecht. "Understanding Emoji Ambiguity in Context: The Role of Text in Emoji-Related Miscommunication." In *Proceedings of the Eleventh International AAAI Conference on Web and Social Media*, edited by Derek Ruths. Palo Alto, CA: AAAI Press, 2017, 152–61.

Miller, Hannah Jean, Jacob Thebault-Spieker, Shuo Chang, Isaac Johnson, Loren Terveen, and Brent Hecht. "'Blissfully Happy' or 'Ready to Fight': Varying Interpretations of Emoji." In *Proceedings of the Tenth International AAAI Conference on Web and Social Media*, edited by Markus Strohmaier and Krishna P. Gummadi. Palo Alto CA: AAAI Press, 2016, 259–68.

Mills, Carol Bishop. "Child's Play or Risky Business? The Development of Teasing Functions and Relational Implications in School-Aged Children." *Journal of Social and Personal Relationships* 35, no. 3 (2018): 287–306.

Mintz, Zoe. "Dontadrian Bruce, 15-Year-Old Student, Suspended after School Says Hand Gesture Represented 'Vice Lord' Gang Sign." *International Business Times*, March 12, 2014.

Mishoe, Margaret, and Michael Montgomery. "The Pragmatics of Multiple Modal Variation in North and South Carolina." *American Speech* 69, no. 1 (1994): 3–29.

Mitchell, Melanie. *Artificial Intelligence: A Guide for Thinking Humans.* New York: Farrar, Straus and Giroux, 2019.

Mole, Beth. "Debunked: The Absurd Story about Smartphones Causing Kids to Sprout Horns." *Ars Technica*, June 21, 2019. https://arstechnica.com/science/2019/06/debunked-the-absurd-story-about-smartphones-causing-kids-to-sprout-horns

Molesworth, Brett R. C., and Dominique Estival. "Miscommunication in General Aviation: The Influence of External Factors on Communication Errors." *Safety Science* 73 (2015): 73–79.

Montgomery, L. M. *Chronicles of Avonlea.* Toronto: McClelland & Stewart, 1912.

Moray, Neville. "Attention in Dichotic Listening: Affective Cues and the Influence of Instructions." *Quarterly Journal of Experimental Psychology* 11, no. 1 (1959): 56–60.

Morgan, Jon. "Radar, Built Here, Detected Pearl Harbor Attack, but . . . Futile Early Warning." *Baltimore Sun*, November 20, 1991. https://www.baltimoresun.com/news/bs-xpm-1991-11-29-1991333114-story.html

Morgioni, Mark. "Defending 'Garbage Language,' the Silly Corporate Terminology That Seriously Works." *Slate*, February 20, 2020. https://slate.com/human-interest/2020/02/garbage-language-business-speak-defense.html

Munro, Kelsey. "How Safe Is Flying? Here's What the Statistics Say." *SBSNews*, July 31, 2018. https://www.sbs.com.au/news/how-safe-is-flying-here-s-what-the-statistics-say

Murphy, Kim. "Driven to Extremes in the Northwest." *Los Angeles Times*, January 11, 1998.

Murray, Rheanna. "School Defends Principal in Controversial 'Gang Sign' Photo." *ABC News*, November 20, 2014.

Musolff, Andreas. "Metaphors: Sources for Intercultural Misunderstanding?" *International Journal of Language and Culture* 1, no. 1 (2014): 42–59.

Naftulin, Donald H., John E. Ware, and Frank A. Donnelly. "The Doctor Fox Lecture: A Paradigm of Educational Seduction." *Journal of Medical Education* 48, no. 7 (1973): 630–35.

Nagy, Emese. "Is Newborn Smiling Really Just a Reflex? Research Is Challenging the Textbooks." *The Conversation*, October 30, 2018. https://theconversation.com/is-newborn-smiling-really-just-a-reflex-research-is-challenging-the-textbooks-105220

Näslund, Shirley. "Tacit Tango: The Social Framework of Screen-Focused Silence in Institutional Telephone Calls." *Journal of Pragmatics* 91 (2016): 60–79.

Nelson, Shellie. "Sen. Grassley Gets Attention for Tweeting 'u kno what.'" *WQAD News*, November 7, 2014. https://wqad.com/2014/11/07/sen-grassley-gets-attention-for-tweeting-u-kno-what

Newell, Jesse. "How Miscommunication Hurt KU Football's Offense in Two Underwhelming Games." *Kansas City Star*, September 9, 2019.

Newlin-Łukowicz, Luiza. "TH-Stopping in New York City: Substrate Effect Turned Ethnic Marker?" *University of Pennsylvania Working Papers in Linguistics* 19, no. 2 (2013): 150–60.

Newman, Andy, and Sarah Maslin Nir. "New York Today: Cucumber Time." *New York Times*, August 19, 2013.

Newton, Elizabeth L. "The Rocky Road from Actions to Intentions." Unpublished doctoral dissertation, Stanford University, Stanford, CA, 1991.

New York Times. "Hyphen or Comma? The History of the Error in the Enrollment Tariff Bill," February 21, 1874.

———. "'In' on 'Inflammable' Found to Mislead Many." January 21, 1947, 10.

———. "Romney Asserts He Underwent 'Brainwashing' on Vietnam Trip." September 5, 1967, 28.

———. "Virginia Tourism Agency Removes Gang Gesture from Ad," August 19, 2007.

———. "Answers from a Court Reporter," June 16, 2010.

Nicolov, Nicolas. "Book Review: Anaphora Resolution." *IEEE Computational Intelligence Bulletin* 2, no. 1 (2003): 31–32.

Nixon, Ron. "Last of a Breed: Postal Workers Who Decipher Bad Addresses." *New York Times*, May 4, 2013, A11.

Northern, Jerry L., and Marion P. Downs. *Hearing in Children*. 5th ed. Philadelphia: Lippincott Williams & Wilkins, 2002.

Norris, Mary. *Between You & Me: Confessions of a Comma Queen*. New York: Norton, 2015.

Nova. "Smartest Machine on Earth." Directed by Michael Bicks, written by Michael Bicks and Julia Cort. *PBS*, February 9, 2011.

O'Brien, Edward J., Anne E. Cook, and Robert F. Lorch Jr. *Inferences during Reading*. Cambridge: Cambridge University Press, 2015.

Ochs, Eleanor. "Misunderstanding Children." In *"Miscommunication" and Problematic Talk*, edited by Nikolas Coupland, Howard Giles, and John M. Wiemann, 44–60. Newbury Park, CA: Sage, 1991.

O'Conner, Patricia T., and Stewart Kellerman. *Origins of the Specious: Myths and Misconceptions of the English Language*. New York: Random House, 2009.

Ohadi, Jonathan, Brandon Brown, Leora Trub, and Lisa Rosenthal. "I Just Text to Say I Love You: Partner Similarity in Texting and Relationship Satisfaction." *Computers in Human Behavior* 78 (2018): 126–32.

Ohlheiser, Abby. "Ken Bone Was a 'Hero.' Now Ken Bone Is 'Bad.' It Was His Destiny as a Human Meme." *Washington Post*, October 14, 2016.

Olasov, Ian. "Offensive Political Dog Whistles: You Know Them When You Hear Them. Or Do You?" *Vox*, November 7, 2016. https://www.vox.com/the-big-idea/2016/11/7/13549154/dog-whistles-campaign-racism

Oppel, Richard A., Jr. "After Debate Gaffe, Perry Vows to Stay in Race." *New York Times*, November 10, 2011.

Padilla, Mariel. "Wait, Is It Nevada, or Nevada?" *New York Times*, February 19, 2020.

Pasquini, Elisabeth S., Kathleen H. Corriveau, Melissa Koenig, and Paul L. Harris. "Preschoolers Monitor the Relative Accuracy of Informants." *Developmental Psychology* 43, no. 5 (2007): 1216–26.

Pastore, Ray. "The Effects of Time-Compressed Instruction and Redundancy on Learning and Learners' Perceptions of Cognitive Load." *Computers & Education* 58, no. 1 (2012): 641–51.

———. "Time-Compressed Instruction: What Compression Speeds Do Learners Prefer." *International Journal of Instructional Technology & Distance Learning* 12, no. 6 (2015): 3–20.

Pastore, Raymond, and Albert D. Ritzhaupt. "Using Time-Compression to Make Multimedia Learning More Efficient: Current Research and Practice." *TechTrends* 59, no. 2 (2015): 66–74.

Patterson, R. Gary. *The Walrus Was Paul: The Great Beatle Death Clues*. New York: Fireside, 1996.

Paxton, Larry J. "'Faster, Better, and Cheaper' at NASA: Lessons Learned in Managing and Accepting Risk." *Acta Astronautica* 61 (2007): 954–63.

Peer, Eyal, and Elisha Babad. "The Doctor Fox Research (1973) Rerevisited: 'Educational Seduction' Ruled Out." *Journal of Educational Psychology* 106, no. 1 (2014): 36–45.

Perfetti, Charles A., Sylvia Beverly, Laura Bell, Kimberly Rodgers, and Robert Faux. "Comprehending Newspaper Headlines." *Journal of Memory and Language* 26, no. 6 (1987): 692–713.

Perlman, Merrill. "How the Word 'Queer' Was Adopted by the LGBTQ Community." *Columbia Journalism Review*, February 22, 2019. https://www.cjr.org/language_corner/queer.php

Peters, Lucia. "Those Emoji Don't Mean What You Think They Do." *Bustle*, April 12, 2016. https://www.bustle.com/articles/154207-emoji-misunderstanding-is-more-common-than-you-think-heres-why-we-dont-all-get-the

Petri, Alexander. "The Best Response to Pointergate, the Worst News Story of 2014?" *Washington Post*, November 14, 2014.

Petty, Richard E., and John T. Cacioppo. "The Elaboration Likelihood Model of Persuasion." In *Advances in Social Psychology*, no. 19, edited by Leonard Berkowitz, 123–205. New York: Springer, 1986.

Pinker, Steven. *The Blank Slate: The Modern Denial of Human Nature*. New York: Penguin Books, 2003.

Ploog, Helmut. *Handwriting Psychology: Personality Reflected in Handwriting*. Bloomington, IN: iUniverse, 2013.

Poesio, Massimo, Roland Stuckardt, and Yannick Versley, eds. *Anaphora Resolution: Algorithms, Resources, and Applications*. Berlin: Springer-Verlag, 2016.

Polizzotti, Mark. *Sympathy for the Traitor: A Translation Manifesto*. Cambridge, MA: MIT Press, 2018.

Popova, Milena. *Sexual Consent*. Cambridge, MA: MIT Press, 2019.

Poria, Soujanya, Devamanyu Hazarika, Navonil Majumder, and Rada Mihalcea. "Beneath the Tip of the Iceberg: Current Challenges and New Directions in Sentiment Analysis Research." Preprint, arXiv:2005.00357, 2020.

Porter, Stephen, Leanne ten Brinke, and Chantal Gustaw. "Dangerous Decisions: The Impact of First Impressions of Trustworthiness on the Evaluation of Legal Evidence and Defendant Culpability." *Psychology, Crime & Law* 16, no. 6 (2010): 477–91.

Prager, Dennis. *The Rational Bible: Exodus, God, Slavery, and Freedom*. Washington, DC: Regnery Faith, 2018.

Pratt, John, and Michelle Miao. "Risk, Populism, and Criminal Law." *New Criminal Law Review* 22, no. 4 (2019): 391–433.

Pronin, Emily, Justin Kruger, Kenneth Savtisky, and Lee Ross. "You Don't Know Me, but I Know You: The Illusion of Asymmetric Insight." *Journal of Personality and Social Psychology* 81, no. 4 (2001): 639–56.

Putnam, Thomas. "The Real Meaning of *Ich Bin ein Berliner*." *The Atlantic*, August 2013.

Quito, Anne. "Why We Can't Stop Using the 'Face with Tears of Joy' Emoji." *Quartz*, October 18, 2019. https://qz.com/1726756/the-psychology-behind -the-most-popular-emoji

Radin, Margaret Jane. *Boilerplate: The Fine Print, Vanishing Rights, and the Rule of Law*. Princeton, NJ: Princeton University Press, 2012.

Radley, David C., Melanie R. Wasserman, Lauren Ew Olsho, Sarah J. Shoemaker, Mark D. Spranca, and Bethany Bradshaw. "Reduction in Medication Errors in Hospitals Due to Adoption of Computerized Provider Order Entry Systems." *Journal of the American Medical Informatics Association* 20, no. 3 (2013): 470–76.

Rai, Roshan, Peter Mitchell, Tasleem Kadar, and Laura Mackenzie. "Adolescent Egocentrism and the Illusion of Transparency: Are Adolescents as Egocentric as We Might Think?" *Current Psychology* 35, no. 3 (2016): 285–94.

Rajadesingan, Ashwin, Reza Zafarani, and Huan Liu. "Sarcasm Detection on Twitter: A Behavioral Modeling Approach." In *Proceedings of the Eighth ACM International Conference on Web Search and Data Mining*, edited by Xueqi Cheng, 97–106. New York: Association for Computing Machinery, 2015.

Rankin, William. "MEDA Investigation Process." *Boeing Aero* 26, no. 2, (2007): 15–21.

Ransohoff, David F., and Richard M. Ransohoff. "Sensationalism in the Media: When Scientists and Journalists May Be Complicit Collaborators." *Effective Clinical Practice* 4, no. 4 (2001): 185–88.

Rauscher, Frances H., Gordon L. Shaw, and Catherine N. Ky. "Music and Spatial Task Performance." *Nature* 365, no. 6447 (1993): 611.

Rawlings, Deborah, Jennifer J. Tieman, Christine Sanderson, Deborah Parker, and Lauren Miller-Lewis. "Never Say Die: Death Euphemisms, Misunderstandings and Their Implications for Practice." *International Journal of Palliative Nursing* 23, no. 7 (2017): 324–30.

Rayner, Keith. "Eye Movements in Reading and Information Processing: 20 Years of Research." *Psychological Bulletin* 124, no. 3 (1998): 372–422.

Rayner, Keith, Elizabeth R. Schotter, Michael E. J. Masson, Mary C. Potter, and Rebecca Treiman. "So Much to Read, so Little Time: How Do We Read, and Can Speed Reading Help?" *Psychological Science in the Public Interest* 17, no. 1 (2016): 4–34.

Read, J. Don. "Detection of Fs in a Single Statement: The Role of Phonetic Recoding." *Memory & Cognition* 11, no. 4 (1983): 390–99.

Reader's Digest. *Laughter Totally Is the Best Medicine.* White Plains, NY: Trusted Media Brands, Inc., 2019.

Reddy, Michael. "The Conduit Metaphor: A Case of Frame Conflict in Our Language about Language." In *Metaphor and Thought*, edited by Andrew Ortony, 284–324. Cambridge: Cambridge University Press, 1979.

Reeves, Carol. *The Language of Science.* London: Routledge, 2005.

Rentoul, John. "The Top 10: Commonly Confused Abbreviations." *The Independent*, February 14, 2020.

Rickford, John R., and Sharese King. "Language and Linguistics on Trial: Hearing Rachel Jeantel (and Other Vernacular Speakers) in the Courtroom and Beyond." *Language* 92, no. 4 (2016): 948–88.

Riordan, Monica A., and Lauren A. Trichtinger. "Overconfidence at the Keyboard: Confidence and Accuracy in Interpreting Affect in E-Mail Exchanges." *Human Communication Research* 43, no. 1 (2017): 1–24.

Roberts, Mary L. *What Soldiers Do: Sex and the American GI in World War II France.* Chicago: University of Chicago Press, 2013.

Roberts, Richard M., and Roger J. Kreuz. "Why Do People Use Figurative Language?" *Psychological Science* 5, no. 3 (1994): 159–63.

Roediger, Henry L., and K. Andrew DeSoto. "Recognizing the Presidents: Was Alexander Hamilton President?" *Psychological Science* 27, no. 5 (2016): 644–50.

Roediger, Henry L., and Kathleen B. McDermott. "Creating False Memories: Remembering Words Not Presented in Lists." *Journal of Experimental Psychology: Learning, Memory, and Cognition* 21, no. 4 (1995): 803–14.

Roghanizad, M. Mahdi, and Vanessa K. Bohns. "Ask in Person: You're Less Persuasive Than You Think over Email." *Journal of Experimental Social Psychology* 69 (2017): 223–26.

Rohan, Tim. "Mets Revamp Sign System after Daniel Murphy's Departure." *New York Times*, March 6, 2016.

Ronson, Jon. "How One Stupid Tweet Blew Up Justine Sacco's Life." *New York Times*, February 12, 2015.

Rosenberg, Marvin. *The Adventures of a Shakespeare Scholar: To Discover Shakespeare's Art.* Newark: University of Delaware Press, 1997.

Rosenblum, Sara, Margalit Samuel, Sharon Zlotnik, Ilana Erikh, and Ilana Schlesinger. "Handwriting as an Objective Tool for Parkinson's Disease Diagnosis." *Journal of Neurology* 260, no. 9 (2013): 2357–61.

Rosenthal, Andrew. "Dukakis Criticizes Bush Tax Cut Plan." *New York Times*, October 25, 1988, A26.

Rosenthal, Jack. "On Language: Misheard, Misread, Misspoken." *New York Times*, August 28, 1988, 20.

Ross, Alex. "Searching for Silence: John Cage's Art of Noise." *The New Yorker*, October 4, 2010.

Ross, Lee, David Greene, and Pamela House. "The 'False Consensus Effect': An Egocentric Bias in Social Perception and Attribution Processes." *Journal of Experimental Social Psychology* 13, no. 3 (1977): 279–301.

RT. "Interpreter of Khrushchev's 'We Will Bury You' Phrase Dies at 81." May 16, 2014. https://www.rt.com/news/159524-sukhodrev-interpreter -khrushchev-cold-war

Rubio-Fernandez, Paula. "Overinformative Speakers Are Cooperative: Revisiting the Gricean Maxim of Quantity." *Cognitive Science* 43, no. 11 (2019): e12797.

Rueb, Emily S. "Cursive Seemed to Go the Way of Quills and Parchment. Now It's Coming Back." *New York Times*, April 13, 2019. https://www.nytimes .com/2019/04/13/education/cursive-writing.html

Sack, Kevin. "Georgia's Governor Seeks Musical Start for Babies." *New York Times*, January 15, 1998.

Sacks, Oliver. *Hallucinations*. New York: Vintage Books, 2012.

———. Mishearings. *New York Times*, June 6, 2015, SR4.

Sales, Ben. "Senator's Speech on 'Cosmopolitan Elites': Anti-Semitic Dog Whistle or Poli-Sci Speak?" *Jewish Telegraphic Agency*, July 19, 2019. https:// www.jta.org/2019/07/19/united-states/a-missouri-senator-gave-a-speech -opposing-a-powerful-upper-class-and-their-cosmopolitan-priorities-um

Salkin, Jeffrey K. *Righteous Gentiles in the Hebrew Bible: Ancient Role Models for Sacred Relationships*. Woodstock, VT: Jewish Lights Publishing, 2008.

Sandomir, Richard. "ESPN Fires Employee over Slur." *New York Times*, February 20, 2012, D2.

Satran, Joe. "Vintage Slang Terms for Being Drunk Are Hilarious a Century Later." *HuffPost*, December 7, 2017. https://www.huffpost.com/entry/vin tage-slang-terms-drunk_n_4268480

Schafer, Sarah. "E-Mail's Impersonal Tone Easily Misunderstood: Conflicts Can Arise from Mistyped, Misperceived Messages." *Washington Post*, November 3, 2000.

Schegloff, Emanuel A., Gail Jefferson, and Harvey Sacks. "The Preference for Self-Correction in the Organization of Repair in Conversation." *Language* 53, no. 2 (1977): 361–82.

Schilling, Govert. *Atlas of Astronomical Discoveries*. New York: Springer Science & Business Media, 2011.

Schmidt, Michael S. "Phillies Are Accused of Stealing Signs Illegally." *New York Times*, May 12, 2010.

———. "Boston Red Sox Used Apple Watches to Steal Signs against Yankees." *New York Times*, September 5, 2017.

Schober, Michael F., and Herbert H. Clark. "Understanding by Addressees and Overhearers." *Cognitive Psychology* 21, no. 2 (1989): 211–32.

Schoenberg, Nara. "OK Sign Is under Siege: How the Squeaky-Clean Hand Gesture Was Twisted by Trolls and Acquired Racist Undertones." *Chicago Tribune*, May 30, 2019.

Schwartz, Martin A. "Legal Writing: Legalese, Please." *Probate & Property* 31 (2017): 55–59.

Seidenberg, Mark. *Language at the Speed of Sight: How We Read, Why So Many Can't, and What Can Be Done about It*. New York: Basic Books, 2017.

Seinfeld. "The Subway." Episode 313. Directed by Tom Cherones. Written by Larry Charles. NBC. January 8, 1992.

Shahar, David, and Mark G. L. Sayers. "Prominent Exostosis Projecting from the Occipital Squama More Substantial and Prevalent in Young Adult Than Older Age Groups." *Scientific Reports* 8, no. 1 (2018): 3354.

Shentiz, Bruce. "New York's Beginnings, Real and Imagined." *New York Times*, December 3, 1999, E35.

Sherzer, Joel. *Speech Play and Verbal Art*. Austin: University of Texas Press, 2002.

Shilling, Janet. "I'll Fight Jargon with Jargon." *New York Times*, August 22, 1987, sec. 1, 27.

Shuchman, Miriam, and Michael S. Wilkes. "Medical Scientists and Health News Reporting: A Case of Miscommunication." *Annals of Internal Medicine* 126, no. 12 (1997): 976–82.

Siegal, Allan M., and William G. Connolly. *The New York Times Manual of Style and Usage*. 5th ed. New York: Three Rivers Press, 2015.

Simon, Herbert A. "Rational Choice and the Structure of the Environment." *Psychological Review* 63, no. 2 (1956): 129–38.

Siwolop, Sana. "Business; Counting Coinages in Jargon." *New York Times*, August 1, 1999, sec. 3, 6.

Smart, Reginald G., Robert E. Mann, and Gina Stoduto. "The Prevalence of Road Rage." *Canadian Journal of Public Health* 94, no. 4 (2003): 247–50.

Smith, Ben. "I'll Take 'Hand Gestures of QAnon' for $1,000." *New York Times*, May 17, 2021, B1.

Smith, W. Donald. "Beyond the 'Bridge on the River Kwai': Labor Mobilization in the Greater East Asia Co-Prosperity Sphere." *International Labor and Working-Class History* 58 (2000): 219–38.

Sohmer, Steve. *Reading Shakespeare's Mind.* Manchester: Manchester University Press, 2017.

Soler, Eva Alcón. "Does Instruction Work for Learning Pragmatics in the EFL Context?" *System 33*, no. 3 (2005): 417–35.

Soniak, Matt. "How Did the Term 'Open a Can of Worms' Originate?" *Mental Floss*, June 28, 2012. https://www.mentalfloss.com/article/31039/how-did -term-open-can-worms-originate

Sparks, Hannah. "Young People Are Growing Horns from Cellphone Use: Study." *New York Post*, June 20, 2019.

Speelman, Craig. "Implicit Expertise: Do We Expect Too Much from Our Experts?" In *Implicit and Explicit Mental Processes,* edited by Kim Kirsner, Craig Speeman, Murray Mabery, Angela O'Brien-Malone, Mike Anderson, and Colin MacLeod, 135–47. Mahwah, NJ: Lawrence Erlbaum Associates, 1998.

Speer, Susan A. "Flirting: A Designedly Ambiguous Action?" *Research on Language and Social Interaction* 50, no. 2 (2017): 128–50.

Stanley-Becker, Isaac. "'Horns' Are Growing on Young People's Skulls. Phone Use Is to Blame, Research Suggests." *Washington Post*, June 25, 2019.

Staub, Adrian, Sophia Dodge, and Andrew L. Cohen. "Failure to Detect Function Word Repetitions and Omissions in Reading: Are Eye Movements to Blame?" *Psychonomic Bulletin & Review* 26, no. 1 (2019): 340–46.

Steinmetz, Katy. "Oxford's 2015 Word of the Year Is This Emoji." *Time*, November 16, 2015.

Stephens, Keri K., Marian L. Houser, and Renee L. Cowan. "RU Able to Meat Me: The Impact of Students' Overly Casual Email Messages to Instructors." *Communication Education* 58, no. 3 (2009): 303–26.

Stephenson, Michael. "Harry Potter Language: The Plain Language Movement and the Case against Abandoning Legalese." *Northern Ireland Legal Quarterly* 68, no. 1 (2017): 85–90.

Stern, Sheldon M. *Averting "The Final Failure": John F. Kennedy and the Secret Cuban Missile Crisis Meetings.* Stanford, CA: Stanford University Press, 2003.

St. John, Warren. "When Information Becomes T.M.I." *New York Times*, September 10, 2006, sec. 9, 9.

Strauss, David. *The Planet Mars: A History of Observation and Discovery.* Tucson: University of Arizona Press, 1996.

Strayer, Lara N. "Ambiguous Laws Do Little to Erase Kiddie Porn." *Temple Political & Civil Rights Law Review* 5 (1995): 169–81.

Sullman, Mark J. M. "The Expression of Anger on the Road." *Safety Science* 72 (2015): 153–59.

Taber, Keith S. "When the Analogy Breaks Down: Modelling the Atom on the Solar System." *Physics Education* 36, no. 3 (2001): 222–26.

———. "Mind Your Language: Metaphor Can Be a Double-Edged Sword." *Physics Education* 40, no. 1 (2005): 11–12.

Tajima, Atsushi. "Fatal Miscommunication: English in Aviation Safety." *World Englishes* 23, no. 3 (2004): 451–70.

Taubman, Willam. *Khrushchev: The Man and His Era.* New York: Norton, 2003.

Tauroza, Steve, and Desmond Allison. "Speech Rates in British English." *Applied Linguistics* 11, no. 1 (1990): 90–105.

Tayler, Michael, and Jane Ogden. "Doctors' Use of Euphemisms and Their Impact on Patients' Beliefs about Health: An Experimental Study of Heart Failure." *Patient Education and Counseling* 57, no. 3 (2005): 321–26.

Taylor, Derrick B. "Professor Fired after Joking That Iran Should Pick U.S. Sites to Bomb." *New York Times,* January 11, 2020.

Taylor, Wilson L. "'Cloze Procedure': A New Tool for Measuring Readability." *Journalism Bulletin* 30, no. 4 (1953): 415–33.

The New Yorker. "Words Came in Marked for Death . . .". April 23, 2012.

Thibodeau, Paul H. "A Moist Crevice for Word Aversion: In Semantics Not Sounds." *PLoS One* 11, no. 4 (2016): e0153686.

Thompson, Alex I. "Wrangling Tips: Entrepreneurial Manipulation in Fast-Food Delivery." *Journal of Contemporary Ethnography* 44, no. 6 (2015): 737–65.

Tierney, John. "The Big City; A Hand Signal to Counteract Road Rage." *New York Times,* March 15, 1999, B1.

Topol, Eric. *The Creative Destruction of Medicine: How the Digital Revolution Will Create Better Health Care.* New York: Basic Books, 2012.

Torres, Ella. "Kathy Zhu, Miss Michigan 2019, Stripped of Her Title over 'Offensive' Social Media Posts." *ABC News,* July 20, 2019. https://abcnews.go.com/US/kathy-zhu-miss-michigan-2019-stripped-title-offensive/story?id=64459158

Toscani, Matteo, Karl R. Gegenfurtner, and Katja Doerschner. "Differences in Illumination Estimation in #thedress." *Journal of Vision* 17, no. 1 (2017): 1–14.

Trubek, Anne. *The History and Uncertain Future of Handwriting.* New York: Bloomsbury, 2016.

Truss, Lynne. *Eats, Shoots & Leaves: The Zero Tolerance Approach to Punctuation.* New York: Avery, 2003.

Tsuchiyama, Jayne. "The Term 'Oriental' Is Outdated, but Is It Racist?" *Los Angeles Times*, June 1, 2016.

Turbow, Jason. *The Baseball Codes: Beanballs, Sign Stealing, and Bench-Clearing Brawls: The Unwritten Rules of America's Pastime.* New York: Pantheon, 2010.

Tyko, Kelly. "'Corona Beer Virus' and 'Beer Coronavirus' Searches Increase as Fears of Outbreak Spread." *USA Today*, February 27, 2020.

USA Today Network. "What Do All Those Hand Signs Mean? Inside the Hidden Language of Baseball and Softball." *Wausau Daily Herald*, April 24, 2018.

Van Hee, Cynthia, Els Lefever, and Véronique Hoste. "We Usually Don't Like Going to the Dentist: Using Common Sense to Detect Irony on Twitter." *Computational Linguistics* 44, no. 4 (2018): 793–832.

Vestergaard, Martin, Mickey T. Kongerslev, Marianne S. Thomsen, Birgit Bork Mathiesen, Catherine J. Harmer, Erik Simonsen, and Kamilla W. Miskowiak. "Women with Borderline Personality Disorder Show Reduced Identification of Emotional Facial Expressions and a Heightened Negativity Bias." *Journal of Personality Disorders* 34, no. 5 (2020): 677–98.

Victor, Daniel. "Oxford Comma Dispute Is Settled as Maine Drivers Get $5 Million." *New York Times*, February 9, 2018.

Vigdor, Neil. "The Houston Astros' Cheating Scandal: Sign-Stealing, Buzzer Intrigue and Tainted Pennants." *New York Times*, July 16, 2020.

Vitello, Paul. "A Ring Tone Meant to Fall on Deaf Ears." *New York Times*, June 12, 2006.

Von Schneidemesser, Luanne. "Soda or Pop?" *Journal of English Linguistics* 24, no. 4 (1996): 270–87.

Walker, C. J., and D. Struzyk. "Evidence for a Social Conduct Moderating Function of Common Gossip." Paper presented to the International Society for the Study of Close Relationships, Saratoga Springs, NY, 1998.

Walker, Gareth. "The Phonetic Constitution of a Turn-Holding Practice: Rush-Throughs in English Talk-in-Interaction." In *Prosody in Interaction*, edited by Dagmar Barth-Weingarten, Elisabeth Reber, and Margret Selting, 51–72. Amsterdam: John Benjamins, 2010.

Walsh, Debra G., and Jay Hewitt. "Giving Men the Come-On: Effect of Eye Contact and Smiling in a Bar Environment." *Perceptual and Motor Skills* 61, no. 3 (1985): 873–74.

Ware, John E., and Reed G. Williams. "The Dr. Fox Effect: A Study of Lecturer Effectiveness and Ratings of Instruction." *Journal of Medical Education*, 50, no. 2 (1975): 149–56.

Warner, Ezra J. *Generals in Gray: Lives of the Confederate Commanders*. Baton Rouge: Louisiana State University Press, 2006.

Wasik, Barbara A., and Annemarie H. Hindman. "Why Wait? The Importance of Wait Time in Developing Young Students' Language and Vocabulary Skills." *The Reading Teacher* 72, no. 3 (2018): 369–78.

Watson, Cecelia. *Semicolon: The Past, Present, and Future of a Misunderstood Mark*. New York: Ecco, 2019.

Waxman, Olivia. "Pronunciation Fail Costs Guy $1 Million Prize on 'Wheel of Fortune.'" *Time*, September 19, 2013.

Weigand, Edda. "Misunderstanding: The Standard Case." *Journal of Pragmatics* 31, no. 6 (1999): 763–85.

Weiser, Benjamin. "At Silk Road Trial, Lawyers Fight to Include Evidence They Call Vital: Emoji." *New York Times*, January 29, 2015, A22.

Wetts, Rachel, and Robb Willer. "Who Is Called by the Dog Whistle? Experimental Evidence That Racial Resentment and Political Ideology Condition Responses to Racially Encoded Messages." *Socius* 5 (2019): 1–20.

Whyte, William H. "Is Anybody Listening?" *Fortune*, September 1950.

Wiener, Anna. *Uncanny Valley: A Memoir*. New York: Farrar, Straus and Giroux, 2020.

Wijewardena, Nilupama, Ramanie Samaratunge, Charmine Härtel, and Andrea Kirk-Brown. "Why Did the Emu Cross the Road? Exploring Employees' Perception and Expectations of Humor in the Australian Workplace." *Australian Journal of Management* 41, no. 3 (2016): 563–84.

Wilford, John N. "Mars Orbiting Craft Presumed Destroyed by Navigation Error." *New York Times*, September 24, 1999, A01.

Wilson, Timothy D., David A. Reinhard, Erin C. Westgate, Daniel T. Gilbert, Nicole Ellerbeck, Cheryl Hahn, Casey L. Brown, and Adi Shaked. "Just Think: The Challenges of the Disengaged Mind." *Science* 345, no. 6192 (2014): 75–77.

Wineapple, Brenda. *The Impeachers: The Trial of Andrew Johnson and the Dream of a Just Nation*. New York: Random House, 2019.

Wogalter, Michael S., Jesseca R. Israel, Soyun Kim, Emily R. Morgan, Kwamoore M. Coleman, and Julianne West. "Hazard Connotation of Fire

Safety Terms." *Proceedings of the Human Factors and Ergonomics Society Annual Meeting* 54, no. 21 (2010): 1837–40. Thousand Oaks CA: SAGE Publications.

Wright, Susan. "The Death of Lady Mondegreen." *Harper's Magazine* 201 (November 1954): 48–51.

Yokley, Eli. "Voters Want Gary Johnson, Jill Stein on the Debate Stage." *Morning Consult*, September 1, 2016. https://morningconsult.com/2016/09/01/voters-want-gary-johnson-jill-stein-debate-stage

Young, Molly. "Garbage Language: Why Do Corporations Speak the Way They Do?" *New York*, February 17, 2020. https://www.thecut.com/2020/02/spread-of-corporate-speak.html

Zimmer, Ben. "Crash Blossoms." *New York Times Magazine*, January 31, 2010, MM14.

Ziomek, Jon. *Collision on Tenerife: The How and Why of the World's Worst Aviation Disaster*. New York: Post Hill Press, 2018.

Zucchino, David, and David S. Cloud. "U.S. Deaths in Drone Strike Due to Miscommunication, Report Says." *Los Angeles Times*, October 14, 2011.

INDEX

ABOUT THE AUTHOR

Roger Kreuz is associate dean in the College of Arts and Sciences and professor of psychology at the University of Memphis. He earned his doctoral degree in cognitive psychology from Princeton University. Kreuz is the author of four other books on language and communication, including *Becoming Fluent: How Cognitive Science Can Help Adults Learn a Foreign Language, Getting Through: The Pleasures and Perils of Cross-Cultural Communication, Changing Minds: How Aging Affects Language and How Language Affects Aging,* and *Irony and Sarcasm.* Translations of these books have appeared in Korean, Russian, Turkish, Japanese, Chinese, and Spanish.